T0245249

CAMBRIDGE LIBRARY COLLECTION

Books of enduring scholarly value

Darwin

Two hundred years after his birth and 150 years after the publication of 'On the Origin of Species', Charles Darwin and his theories are still the focus of worldwide attention. This series offers not only works by Darwin, but also the writings of his mentors in Cambridge and elsewhere, and a survey of the impassioned scientific, philosophical and theological debates sparked by his 'dangerous idea'.

The Different Forms of Flowers on Plants of the Same Species

After the publication of *On the Origin of Species* in 1859 Darwin became fascinated with the potential for botanical experiments to provide evidence for the process of evolution. First published in 1877, this volume is based on a series of papers concerning heterostylous plants (species which produce different types of flowers) originally published in the *Journal of the Proceedings of the Linnaen Society* in 1862. Linnaeus had divided the sexual relations of flowers into four groups, which Darwin uses as the structure for this volume. Darwin examines in detail plants which produce different flower forms, presenting his conclusions in terms of adaptive evolution and so providing the first functional interpretation of heterostyly. He demonstrates that these plants are adapted for cross-fertilisation, not self-fertilisation as was widely believed. The concepts which Darwin introduces in this volume continue to provide the basis for research into plant reproductive biology.

Cambridge University Press has long been a pioneer in the reissuing of out-of-print titles from its own backlist, producing digital reprints of books that are still sought after by scholars and students but could not be reprinted economically using traditional technology. The Cambridge Library Collection extends this activity to a wider range of books which are still of importance to researchers and professionals, either for the source material they contain, or as landmarks in the history of their academic discipline.

Drawing from the world-renowned collections in the Cambridge University Library, and guided by the advice of experts in each subject area, Cambridge University Press is using state-of-the-art scanning machines in its own Printing House to capture the content of each book selected for inclusion. The files are processed to give a consistently clear, crisp image, and the books finished to the high quality standard for which the Press is recognised around the world. The latest print-on-demand technology ensures that the books will remain available indefinitely, and that orders for single or multiple copies can quickly be supplied.

The Cambridge Library Collection will bring back to life books of enduring scholarly value (including out-of-copyright works originally issued by other publishers) across a wide range of disciplines in the humanities and social sciences and in science and technology.

The Different Forms of Flowers on Plants of the Same Species

Charles Darwin

CAMBRIDGE UNIVERSITY PRESS

Cambridge, New York, Melbourne, Madrid, Cape Town, Singapore,
São Paolo, Delhi, Dubai, Tokyo, Mexico City

Published in the United States of America by Cambridge University Press, New York

www.cambridge.org
Information on this title: www.cambridge.org/9781108018272

© in this compilation Cambridge University Press 2010

This edition first published 1877
This digitally printed version 2010

ISBN 978-1-108-01827-2 Paperback

THE

DIFFERENT FORMS OF FLOWERS

ON

PLANTS OF THE SAME SPECIES.

By CHARLES DARWIN, M.A., F.R.S.

WITH ILLUSTRATIONS.

LONDON:
JOHN MURRAY, ALBEMARLE STREET.
1877.

The right of Translation is reserved.

BY THE SAME AUTHOR.

ON THE ORIGIN OF SPECIES BY MEANS OF NATURAL SELECTION; or, THE PRESERVATION OF FAVOURED RACES IN THE STRUGGLE FOR LIFE. Sixth Edition. *Sixteenth Thousand.* MURRAY.

THE DESCENT OF MAN, AND SELECTION IN RELATION TO SEX. With Illustrations. Second Edition, revised and augmented. *Twelfth Thousand.* MURRAY.

THE VARIATION OF ANIMALS AND PLANTS UNDER DOMESTICATION. With Illustrations. Second Edition, revised. *Fourth Thousand.* 2 vols. MURRAY.

THE EXPRESSION OF THE EMOTIONS IN MAN AND ANIMALS. With Photographic and other Illustrations. *Ninth Thousand.* MURRAY.

THE VARIOUS CONTRIVANCES BY WHICH ORCHIDS ARE FERTILISED BY INSECTS. *Second Edition, revised.* With Woodcuts. MURRAY.

INSECTIVOROUS PLANTS. With Illustrations. *Third Thousand.* MURRAY.

THE MOVEMENTS AND HABITS OF CLIMBING PLANTS. With Illustrations. *Second Edition.* MURRAY.

THE EFFECTS OF CROSS AND SELF-FERTILISATION IN THE VEGETABLE KINGDOM. *Second Thousand.* MURRAY.

THE DIFFERENT FORMS OF FLOWERS ON PLANTS OF THE SAME SPECIES. MURRAY.

NATURALIST'S VOYAGE ROUND THE WORLD; or, A JOURNAL OF RESEARCHES INTO THE NATURAL HISTORY AND GEOLOGY OF THE COUNTRIES VISITED during the Voyage of H.M.S. 'Beagle,' under the Command of Captain FITZROY, R.N. *Thirteenth Thousand.* MURRAY.

ON THE STRUCTURE AND DISTRIBUTION OF CORAL REEFS. *Second Edition.* SMITH, ELDER & Co.

GEOLOGICAL OBSERVATIONS ON THE VOLCANIC ISLANDS AND PARTS OF SOUTH AMERICA, visited during the Voyage of H.M.S. 'Beagle.' *Second Edition.* SMITH, ELDER & Co.

A MONOGRAPH OF THE CIRRIPEDIA. With numerous Illustrations. 2 vols. RAY SOCIETY. HARDWICKE.

A MONOGRAPH OF THE FOSSIL LEPADIDÆ OR PEDUNCULATED CIRRIPEDS OF GREAT BRITAIN. PALÆONTOGRAPHICAL SOCIETY.

A MONOGRAPH OF THE FOSSIL BALANIDÆ AND VERRUCIDÆ OF GREAT BRITAIN. PALÆONTOGRAPHICAL SOCIETY.

FACTS AND ARGUMENTS FOR DARWIN. By FRITZ MÜLLER. From the German, with Additions by the Author. Translated by W. S. DALLAS, F.L.S. With Illustrations. MURRAY.

LONDON: PRINTED BY WILLIAM CLOWES AND SONS, STAMFORD STREET AND CHARING CROSS.

TO

PROFESSOR ASA GRAY

𝕿𝖍𝖎𝖘 𝖁𝖔𝖑𝖚𝖒𝖊 𝖎𝖘 𝕯𝖊𝖉𝖎𝖈𝖆𝖙𝖊𝖉

BY THE AUTHOR

AS A SMALL TRIBUTE OF RESPECT AND
AFFECTION.

CONTENTS.

CHAPTER III.

HETEROSTYLED DIMORPHIC PLANTS—*continued.*

CHAPTER IV.

HETEROSTYLED TRIMORPHIC PLANTS.

CHAPTER V.

ILLEGITIMATE OFFSPRING OF HETEROSTYLED PLANTS.

CHAPTER VI.

CONCLUDING REMARKS ON HETEROSTYLED PLANTS.

CHAPTER VII.

POLYGAMOUS, DIŒCIOUS, AND GYNO-DIŒCIOUS PLANTS.

CHAPTER VIII.

CLEISTOGAMIC FLOWERS.

DIFFERENT FORMS OF FLOWERS

ON

PLANTS OF THE SAME SPECIES.

—◆—

INTRODUCTION.

THE subject of the present volume, namely the differently formed flowers normally produced by certain kinds of plants, either on the same stock or on distinct stocks, ought to have been treated by a professed botanist, to which distinction I can lay no claim. As far as the sexual relations of flowers are concerned, Linnæus long ago divided them into hermaphrodite, monœcious, diœcious, and polygamous species. This fundamental distinction, with the aid of several subdivisions in each of the four classes, will serve my purpose; but the classification is artificial, and the groups often pass into one another.

The hermaphrodite class contains two interesting sub-groups, namely, heterostyled and cleistogamic plants; but there are several other less important subdivisions, presently to be given, in which flowers differing in various ways from one another are produced by the same species.

Some plants were described by me several years ago, in a series of papers read before the Linnean Society,*

---

* "On the Two Forms or Dimorphic Condition in the Species of Primula, and on their remarkable Sexual Relations." 'Journal

the individuals of which exist under two or three
forms, differing in the length of their pistils and
stamens and in other respects. They were called by
me dimorphic and trimorphic, but have since been
better named by Hildebrand, heterostyled.* As I
have many still unpublished observations with respect
to these plants, it has seemed to me advisable to re-
publish my former papers in a connected and cor-
rected form, together with the new matter. It will be
shown that these heterostyled plants are adapted for
reciprocal fertilisation; so that the two or three forms,
though all are hermaphrodites, are related to one
another almost like the males and females of ordinary
unisexual animals. I will also give a full abstract of
such observations as have been published since the
appearance of my papers; but only those cases will be
noticed, with respect to which the evidence seems fairly
satisfactory. Some plants have been supposed to be
heterostyled merely from their pistils and stamens
varying greatly in length, and I have been myself
more than once thus deceived. With some species the

of the Proceedings of the Linnean
Society,' vol. vi. 1862, p. 77.
   "On the Existence of Two
Forms, and on their Reciprocal
Sexual Relation, in several Species
of the Genus Linum." Ibid. vol.
vii. 1863, p. 69.
   "On the Sexual Relations of the
Three Forms of *Lythrum salicaria*.'
Ibid. vol. viii. 1864, p. 169.
   "On the Character and Hybrid-
like Nature of the Offspring from
the Illegitimate Unions of Dimor-
phic and Trimorphic Plants."
Ibid. vol. x. 1868, p. 393.
   "On the Specific Differences
between *Primula veris*, Brit. Fl.
(var. *officinalis*, Linn.), *P. vulgaris*,
Brit. Fl. (var. *acaulis*, Linn.), and

P. elatior, Jacq.; and on the
Hybrid Nature of the Common
Oxlip. With Supplementary Remarks on Naturally Produced Hybrids in the Genus Verbascum."
Ibid. vol. x. 1868, p. 437.
   * The term "heterostyled" does
not express all the differences between the forms; but this is a
failure common in many cases.
As the term has been adopted by
writers in various countries, I am
unwilling to change it for that of
heterogone or *heterogonous*, though
this has been proposed by so high
an authority as Prof. Asa Gray:
see the 'American Naturalist,'
Jan. 1877, p. 42.

pistil continues growing for a long time, so that if old and young flowers are compared they might be thought to be heterostyled. Again, a species tending to become diœcious, with the stamens reduced in some individuals and with the pistils in others, often presents a deceptive appearance. Unless it be proved that one form is fully fertile only when it is fertilised with pollen from another form, we have not complete evidence that the species is heterostyled. But when the pistils and stamens differ in length in two or three sets of individuals, and this is accompanied by a difference in the size of the pollen-grains or in the state of the stigma, we may infer with much safety that the species is heterostyled. I have, however, occasionally trusted to a difference between the two forms in the length of the pistil alone, or in the length of the stigma together with its more or less papillose condition ; and in one instance differences of this kind have been proved by trials made on the fertility of the two forms, to be sufficient evidence.

The second sub-group above referred to consists of hermaphrodite plants, which bear two kinds of flowers —the one perfect and fully expanded—the other minute, completely closed, with the petals rudimentary, often with some of the anthers aborted, and the remaining ones together with the stigmas much reduced in size ; yet these flowers are perfectly fertile. They have been called by Dr. Kuhn* cleistogamic, and they

---

* 'Botanische Zeitung,' 1867, p. 65. Several plants are known occasionally to produce flowers destitute of a corolla; but they belong to a different class of cases from cleistogamic flowers. This deficiency seems to result from the conditions to which the plants have been subjected, and partakes of the nature of a monstrosity. All the flowers on the same plant are commonly affected in the same manner. Such cases, though they have sometimes been ranked as cleistogamic, do not come within our present scope : see Dr. Maxwell Masters, 'Vegetable Teratology,' 1869, p. 403.

will be described in the last chapter of this volume.
They are manifestly adapted for self-fertilisation, which
is effected at the cost of a wonderfully small expenditure of pollen; whilst the perfect flowers produced by
the same plant are capable of cross-fertilisation. Certain aquatic species, when they flower beneath the
water, keep their corollas closed, apparently to protect
their pollen; they might therefore be called cleistogamic, but for reasons assigned in the proper place are
not included in the present sub-group. Several cleistogamic species, as we shall hereafter see, bury their
ovaries or young capsules in the ground; but some few
other plants behave in the same manner; and, as they
do not bury all their flowers, they might have formed
a small separate subdivision.

Another interesting subdivision consists of certain
plants, discovered by H. Müller, some individuals
of which bear conspicuous flowers adapted for cross-
fertilisation by the aid of insects, and others much
smaller and less conspicuous flowers, which have often
been slightly modified so as to ensure self-fertilisation.
*Lysimachia vulgaris, Euphrasia officinalis, Rhinanthus
crista-galli,* and *Viola tricolor* come under this head.[*]
The smaller and less conspicuous flowers are not closed,
but as far as the purpose which they serve is concerned, namely, the assured propagation of the species,
they approach in nature cleistogamic flowers; but they
differ from them by the two kinds being produced on
distinct plants.

With many plants, the flowers towards the outside of
the inflorescence are much larger and more conspicuous than the central ones. As I shall not have occa-

* H. Müller, 'Nature,' Sept. 25,
1873 (vol. viii.), p. 433, and Nov.
20, 1873 (vol. ix.), p. 44. Also
'Die Befruchtung der Blumen,'
&c., 1873, p. 294.

sion to refer to plants of this kind in the following chapters, I will here give a few details respecting them. It is familiar to every one that the ray-florets of the Compositæ often differ remarkably from the others ; and so it is with the outer flowers of many Umbelliferæ, some Cruciferæ and a few other families. Several species of Hydrangea and Viburnum offer striking instances of the same fact. The Rubiaceous genus Mussænda presents a very curious appearance from some of the flowers having the tip of one of the sepals developed into a large petal-like expansion, coloured either white or purple. The outer flowers in several Acanthaceous genera are large and conspicuous but sterile ; the next in order are smaller, open, moderately fertile and capable of cross-fertilisation ; whilst the central ones are cleistogamic, being still smaller, closed and highly fertile ; so that here the inflorescence consists of three kinds of flowers.* From what we know in other cases of the use of the corolla, coloured bracteæ, &c., and from what H. Müller has observed† on the frequency of the visits of insects to the flower-heads of the Umbelliferæ and Compositæ being largely determined by their conspicuousness, there can be no doubt that the increased size of the corolla of the outer flowers, the inner ones being in all the above cases small, serves to attract insects. The result is that cross-fertilisation is thus favoured. Most flowers wither soon after being fertilised, but Hildebrand states‡ that the ray-florets of the Compositæ last for a long time, until all those on the disc are impregnated ; and this clearly shows the use of the former. The ray-florets,

---

* J. Scott, 'Journal of Botany,' London, new series, vol. i. 1872, pp. 161-164.
† 'Die Befruchtung der Blu-
men,' pp. 108, 412.
‡ See his interesting memoir, 'Ueber die Geschlechtsverhältnisse bei den Compositen,' 1869, p. 92.

however, are of service in another and very different
manner, namely, by folding inwards at night and
during cold rainy weather, so as to protect the florets
of the disc.* Moreover they often contain matter
which is excessively poisonous to insects, as may be
seen in the use of flea-powder, and in the case of
Pyrethrum, M. Belhomme has shown that the ray-
florets are more poisonous than the disc-florets in the
ratio of about three to two. We may therefore believe
that the ray-florets are useful in protecting the flowers
from being gnawed by insects.†

It is a well-known yet remarkable fact that the circumferential flowers of many of the foregoing plants
have both their male and female reproductive organs
aborted, as with the Hydrangea, Viburnum and certain
Compositæ; or the male organs alone are aborted, as
in many Compositæ. Between the sexless, female and
hermaphrodite states of these latter flowers, the finest
gradations may be traced, as Hildebrand has shown.‡
He also shows that there is a close relation between
the size of the corolla in the ray-florets and the degree
of abortion in their reproductive organs. As we have
good reason to believe that these florets are highly
serviceable to the plants which possess them, more
especially by rendering the flower-heads conspicuous

---

* Kerner clearly shows that
this is the case : ' Die Schutzmittel
des Pollens,' 1873, p. 28.

† ' Gardener's Chronicle,' 1861,
p. 1067. Lindley, ' Vegetable
Kingdom,' on Chrysanthemum,
1853, p. 706. Kerner in his interesting essay (' Die Schutzmittel
der Blüthen gegen unberufene
Gäste,' 1875, p. 19) insists that
the petals of most plants contain
matter which is offensive to insects, so that they are seldom
gnawed, and thus the organs of
fructification are protected. My
grandfather in 1790 (' Loves of
the Plants,' canto iii. note to lines
184, 188) remarks that " The
flowers or petals of plants are
perhaps in general more acrid
than their leaves; hence they are
much seldomer eaten by insects."

‡ ' Ueber die Geschlechtsverhältnisse bei den Compositen,'
1869, pp. 78–91.

to insects, it is a natural inference that their corollas have been increased in size for this special purpose; and that their development has subsequently led, through the principle of compensation or balancement, to the more or less complete reduction of the reproductive organs. But an opposite view may be maintained, namely, that the reproductive organs first began to fail, as often happens under cultivation,* and, as a consequence, the corolla became, through compensation, more highly developed. This view, however, is not probable, for when hermaphrodite plants become diœcious or gyno-diœcious—that is, are converted into hermaphrodites and females—the corolla of the female seems to be almost invariably reduced in size in consequence of the abortion of the male organs. The difference in the result in these two classes of cases, may perhaps be accounted for by the matter saved through the abortion of the male organs in the females of gyno-diœcious and diœcious plants being directed (as we shall see in a future chapter) to the formation of an increased supply of seeds; whilst in the case of the exterior florets and flowers of the plants which we are here considering, such matter is expended in the development of a conspicuous corolla. Whether in the present class of cases the corolla was first affected, as seems to me the more probable view, or the reproductive organs first failed, their states of development are now firmly correlated. We see this well illustrated in Hydrangea and Viburnum; for when these plants are cultivated, the corollas of both the interior and exterior flowers become largely developed, and their reproductive organs are aborted.

---

* I have discussed this subject in my 'Variation of Animals and Plants under Domestication,' chap. xviii. 2nd edit. vol. ii. pp. 152, 156.

There is a closely analogous subdivision of plants, including the genus Muscari (or Feather Hyacinth) and the allied Bellevalia, which bear both perfect flowers and closed bud-like bodies that never expand. The latter resemble in this respect cleistogamic flowers, but differ widely from them in being sterile and conspicuous. Not only the aborted flower-buds and their peduncles (which are elongated apparently through the principle of compensation) are brightly coloured, but so is the upper part of the spike—all, no doubt, for the sake of guiding insects to the inconspicuous perfect flowers. From such cases as these we may pass on to certain Labiatæ, for instance, *Salvia Horminum*, in which (as I hear from Mr. Thiselton Dyer) the upper bracts are enlarged and brightly coloured, no doubt for the same purpose as before, with the flowers suppressed.

In the Carrot and some allied Umbelliferæ, the central flower has its petals somewhat enlarged, and these are of a dark purplish-red tint; but it cannot be supposed that this one small flower makes the large white umbel at all more conspicuous to insects. The central flowers are said * to be neuter or sterile, but I obtained by artificial fertilisation a seed (fruit) apparently perfect from one such flower. Occasionally two or three of the flowers next to the central one are similarly characterised; and according to Vaucher† " cette singuliere dégénération s'étend quelquefois à l'ombelle entière." That the modified central flower is of no functional importance to the plant is almost certain. It may perhaps be a remnant of a former and ancient condition of the species, when one flower alone, the

---

* 'The English Flora,' by Sir J. E. Smith, 1824, vol. ii. p. 39.
† 'Hist. Phys. des Plantes d'Europe,' 1841, tom. ii. p. 614. On the Echinophora, p. 627.

central one, was female and yielded seeds, as in the umbelliferous genus Echinophora. There is nothing surprising in the central flower tending to retain its former condition longer than the others; for when irregular flowers become regular or peloric, they are apt to be central; and such peloric flowers apparently owe their origin either to arrested development—that is, to the preservation of an early stage of development—or to reversion. Central and perfectly developed flowers in not a few plants in their normal condition (for instance, the common Rue and Adoxa) differ slightly in structure, as in the number of the parts, from the other flowers on the same plant. All such cases seem connected with the fact of the bud which stands at the end of the shoot being better nourished than the others, as it receives the most sap.*

The cases hitherto mentioned relate to hermaphrodite species which bear differently constructed flowers; but there are some plants that produce differently formed seeds, of which Dr. Kuhn has given a list.† With the Umbelliferæ and Compositæ, the flowers that produce these seeds likewise differ, and the differences in the structure of the seeds are of a very important nature. The causes which have led to differences in the seeds on the same plant are not known; and it is very doubtful whether they subserve any special end.

We now come to our second Class, that of monœcious species, or those which have their sexes separated but borne on the same plant. The flowers necessarily differ, but when those of one sex include rudiments

* This whole subject, including pelorism, has been discussed, and references given, in my ' Variation of Animals and Plants under Domestication,' chap. xxvi. 2nd edit. vol. ii. p. 338.
† ' Bot. Zeitung,' 1867, p. 67.

of the other sex, the difference between the two kinds is usually not great. When the difference is great, as we see in catkin-bearing plants, this depends largely on many of the species in this, as well as in the next or diœcious class, being fertilised by the aid of the wind;* for the male flowers have in this case to produce a surprising amount of incoherent pollen. Some few monœcious plants consist of two bodies of individuals, with their flowers differing in function, though not in structure; for certain individuals mature their pollen before the female flowers on the same plant are ready for fertilisation, and are called proterandrous; whilst conversely other individuals, called proterogynous, have their stigmas mature before their pollen is ready. The purpose of this curious functional difference obviously is to favour the cross-fertilisation of distinct plants. A case of this kind was first observed by Delpino in the Walnut (*Juglans regia*), and has since been observed with the common Nut (*Corylus avellana*). I may add that according to H. Müller the individuals of some few hermaphrodite plants differ in a like manner; some being proterandrous and others proterogynous.† On cultivated trees of the Walnut and Mulberry, the male flowers have been observed to abort on certain individuals‡, which have thus been converted into females; but whether there are any species in a state of nature which co-exist as monœcious and female individuals, I do not know.

The third Class consists of diœcious species, and the

---

* Delpino, 'Studi sopra uno Lignaggio Anemofilo.' Firenze, 1871.

† Delpino, 'Ult. Osservazioni sulla Dicogamia,' part ii. fasc. ii. p. 337. Mr. Wetterhan and H. Müller on Corylus, 'Nature,' vol. xi. p. 507, and 1875, p. 26. On proterandrous and proterogynous hermaphrodite individuals of the same species, see H. Müller, 'Die Befruchtung,' &c., pp. 285, 339.

‡ 'Gardener's Chron.' 1847, pp. 541, 558.

remarks made under the last class with respect to the amount of difference between the male and female flowers are here applicable. It is at present an inexplicable fact that with some diœcious plants, of which the Restiaceæ of Australia and the Cape of Good Hope offer the most striking instance, the differentiation of the sexes has affected the whole plant to such an extent (as I hear from Mr. Thiselton Dyer) that Mr. Bentham and Professor Oliver have often found it impossible to match the male and female specimens of the same species. In my seventh chapter some observations will be given on the gradual conversion of heterostyled and of ordinary hermaphrodite plants into diœcious or sub-diœcious species.

The fourth and last Class consists of the plants which were called polygamous by Linnæus; but it appears to me that it would be convenient to confine this term to the species which co-exist as hermaphrodites, males and females; and to give new names to several other combinations of the sexes—a plan which I shall here follow. Polygamous plants, in this confined sense of the term, may be divided into two sub-groups, according as the three sexual forms are found on the same individual or on distinct individuals. Of this latter or trioicous sub-group, the common Ash (*Fraxinus excelsior*) offers a good instance: thus, I examined during the spring and autumn fifteen trees growing in the same field; and of these, eight produced male flowers alone, and in the autumn not a single seed; four produced only female flowers, which set an abundance of seeds; three were hermaphrodites, which had a different aspect from the other trees whilst in flower, and two of them produced nearly as many seeds as the female trees, whilst the third produced none, so that it was in function a male. The separation of the sexes, however, is not complete in the Ash; for the female flowers include stamens, which drop off at an early period, and their anthers, which never open or dehisce, generally contain pulpy matter instead of pollen. On some female trees, however, I found a few anthers containing pollen-grains apparently sound. On the male trees most of the flowers include pistils, but these likewise drop off at an early period; and the ovules, which ultimately abort, are very small compared with those in female flowers of the same age.

Of the other or monoicous sub-group of polygamous plants, or those which bear hermaphrodite, male and female flowers on the same individual, the common Maple (*Acer campestre*) offers a good instance; but Lecoq states * that some trees are truly diœcious, and this shows how easily one state passes into another.

A considerable number of plants generally ranked as polygamous exist under only two forms, namely, as hermaphrodites and females; and these may be called gyno-diœcious, of which the common Thyme offers a good example. In my seventh chapter I shall give some observations on plants of this nature. Other species, for instance several kinds of Atriplex, bear on the same plant hermaphrodite and female flowers; and these might be called gyno-monœcious, if a name were desirable for them.

Again there are plants which produce hermaphrodite and male flowers on the same individual, for instance, some species of Galium, Veratrum, &c.; and these might be called andro-monœcious. If there exist plants, the individuals of which consist of hermaphrodites and males, these might be distinguished

---

* ' Géographie Botanique,' tom. v. p. 367.

as andro-diœcious. But, after making inquiries from several botanists, I can hear of no such cases. Lecoq, however, states,* but without entering into full details, that some plants of *Caltha palustris* produce only male flowers, and that these live mingled with the hermaphrodites. The rarity of such cases as this last one is remarkable, as the presence of hermaphrodite and male flowers on the same individual is not an unusual occurrence; it would appear as if nature did not think it worth while to devote a distinct individual to the production of pollen, excepting when this was indispensably necessary, as in the case of diœcious species.

I have now finished my brief sketch of the several cases, as far as known to me, in which flowers differing in structure or in function are produced by the same species of plant. Full details will be given in the following chapters with respect to many of these plants. I will begin with the heterostyled, then pass on to certain diœcious, sub-diœcious, and polygamous species, and end with the cleistogamic. For the convenience of the reader, and to save space, the less important cases and details have been printed in smaller type.

I cannot close this Introduction without expressing my warm thanks to Dr. Hooker for supplying me with specimens and for other aid; and to Mr. Thiselton Dyer and Professor Oliver for giving me much information and other assistance. Professor Asa Gray, also, has uniformly aided me in many ways. To Fritz Müller of St. Catharina, in Brazil, I am indebted for many dried flowers of heterostyled plants, often accompanied with valuable notes.

---

* 'Géographie Botanique,' tom. iv. p. 488.

CHAPTER I.

IT has long been known to botanists that the common
Cowslip (*Primula veris*, Brit. Flora, var. *officinalis*,
Lin.) exists under two forms, about equally numerous,
which obviously differ from each other in the length
of their pistils and stamens.* This difference has
hitherto been looked at as a case of mere variability, but this view, as we shall presently see, is far
from the true one. Florists who cultivate the Polyanthus and Auricula have long been aware of the two
kinds of flowers, and they call the plants which display the globular stigma at the mouth of the corolla,
"pin-headed" or "pin-eyed," and those which display
the anthers, "thrum-eyed."† I will designate the two
forms as the long-styled and short-styled.

The pistil in the long-styled form is almost exactly
twice as long as that of the short-styled. The stigma

---

* This fact, according to von
Mohl ('Bot. Zeitung,' 1863, p. 326)
was first observed by Persoon in
the year 1794.
† In Johnson's Dictionary,
thrum is said to be the ends of
weavers' threads; and I suppose

that some weaver who cultivated
the polyanthus invented this name,
from being struck with some degree
of resemblance between the cluster
of anthers in the mouth of the
corolla and the ends of his
threads.

stands in the mouth of the corolla or projects just above it, and is thus externally visible. It stands high above the anthers, which are situated halfway down the tube and cannot be easily seen. In the short-styled form the anthers are attached near the mouth of the tube, and therefore stand above the stigma, which is seated in about the middle of the tubular corolla. The corolla itself is of a different

Fig. 1.

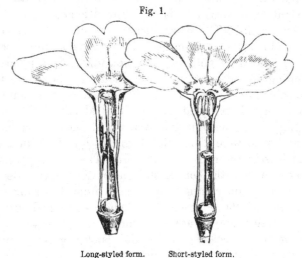

Long-styled form.   Short-styled form.

PRIMULA VERIS.

shape in the two forms; the throat or expanded portion above the attachment of the anthers being much longer in the long-styled than in the short-styled form. Village children notice this difference, as they can best make necklaces by threading and slipping the corollas of the long-styled flowers into one another. But there are much more important differences. The stigma in the long-styled form

is globular; in the short-styled it is depressed on
the summit, so that the longitudinal axis of the
former is sometimes nearly double that of the latter.
Although somewhat variable in shape, one difference
is persistent, namely, in roughness: in some specimens carefully compared, the papillæ which render
the stigma rough were in the long-styled form from
twice to thrice as long as in the short-styled. The
anthers do not differ in size in the two forms, which
I mention because this is the case with some heterostyled plants. The most remarkable difference is in
the pollen-grains. I measured with the micrometer
many specimens, both dry and wet, taken from plants
growing in different situations, and always found a
palpable difference. The grains distended with water
from the short-styled flowers were about ·038 mm.
($\frac{10-11}{7000}$ of an inch) in diameter, whilst those from the
long-styled were about ·0254 mm. ($\frac{7}{7000}$ of an inch),
which is in the ratio of 100 to 67. The pollen-grains
therefore from the longer stamens of the short-styled
form are plainly larger than those from the shorter
stamens of the long-styled. When examined dry,
the smaller grains are seen under a low power to
be more transparent than the larger grains, and
apparently in a greater degree than can be accounted for by their less diameter. There is also a
difference in shape, the grains from the short-styled
plants being nearly spherical, those from the long-
styled being oblong with the angles rounded; this
difference disappears when the grains are distended
with water. The long-styled plants generally tend
to flower a little before the short-styled: for instance,
I had twelve plants of each form growing in separate
pots and treated in every respect alike; and at the
time when only a single short-styled plant was in

flower, seven of the long-styled had expanded their flowers.

We shall, also, presently see that the short-styled plants produce more seed than the long-styled. It is remarkable, according to Prof. Oliver,* that the ovules in the unexpanded and unimpregnated flowers of the latter are considerably larger than those of the short-styled flowers; and this I suppose is connected with the long-styled flowers producing fewer seeds, so that the ovules have more space and nourishment for rapid development.

To sum up the differences:—The long-styled plants have a much longer pistil, with a globular and much rougher stigma, standing high above the anthers. The stamens are short; the grains of pollen smaller and oblong in shape. The upper half of the tube of the corolla is more expanded. The number of seeds produced is smaller and the ovules larger. The plants tend to flower first.

The short-styled plants have a short pistil, half the length of the tube of the corolla, with a smooth depressed stigma standing beneath the anthers. The stamens are long; the grains of pollen are spherical and larger. The tube of the corolla is of uniform diameter except close to the upper end. The number of seeds produced is larger.

I have examined a large number of flowers; and though the shape of the stigma and the length of the pistil both vary, especially in the short-styled form, I have never met with any transitional states between the two forms in plants growing in a state of nature. There is never the slightest doubt under which form a plant ought to be classed. The two kinds of flowers are

---

* 'Nat. Hist. Review,' July 1862, p. 237.

C

never found on the same individual plant. I marked
many Cowslips and Primroses, and on the following
year all retained the same character, as did some in my
garden which flowered out of their proper season in the
autumn. Mr. W. Wooler, of Darlington, however, informs us that he has seen early blossoms on the Polyanthus,* which were not long-styled, but became so later
in the season. Possibly in this case the pistils may not
have been fully developed during the early spring. An
excellent proof of the permanence of the two forms may
be seen in nursery-gardens, where choice varieties of
the Polyanthus are propagated by division; and I found
whole beds of several varieties, each consisting exclusively of the one or the other form. The two forms exist
in the wild state in about equal numbers: I collected
522 umbels from plants growing in several stations,
taking a single umbel from each plant; and 241 were
long-styled, and 281 short-styled. No difference in
tint or size could be perceived in the two great masses
of flowers.

We shall presently see that most of the species of
Primula exist under two analogous forms; and it may
be asked what is the meaning of the above-described
important differences in their structure? The question seems well worthy of careful investigation, and I
will give my observations on the cowslip in detail.
The first idea which naturally occurred to me was,
that this species was tending towards a diœcious
condition; that the long-styled plants, with their
longer pistils, rougher stigmas, and smaller pollen-
grains, were more feminine in nature, and would produce more seed;—that the short-styled plants, with
their shorter pistils, longer stamens and larger pol-

---

* I have proved by numerous
experiments, hereafter to be given,
that the Polyanthus is a variety
of *Primula veris.*

len-grains, were more masculine in nature. Accordingly, in 1860, I marked a few cowslips of both forms growing in my garden, and others growing in an open field, and others in a shady wood, and gathered and weighed the seed. In all the lots the short-styled plants yielded, contrary to my expectation, most seed. Taking the lots together, the following is the result:—

TABLE 1.

—	Number of Plants.	Number of Umbels produced.	Number of Capsules produced.	Weight of Seed in grains.
Short-styled cowslips . . .	9	33	199	83
Long-styled cowslips . . .	13	51	261	91

If we compare the weight from an equal number of plants, and from an equal number of umbels, and from an equal number of capsules of the two forms, we get the following results:—

TABLE 2.

—	Number of Plants.	Weight of Seed in grains.	Number of Umbels.	Weight of Seed.	Number of Capsules.	Weight of Seed in grains.
Short-styled cowslips .	10	92	100	251	100	41
Long-styled cowslips .	10	70	100	178	100	34

So that, by all these standards of comparison, the short-styled form is the more fertile; if we take the number of umbels (which is the fairest standard, for large and small plants are thus equalised), the short-styled plants produce more seed than the long-styled, in the proportion of nearly four to three.

In 1861 the trial was made in a fuller and fairer

C 2

manner. A number of wild plants had been transplanted during the previous autumn into a large bed
in my garden, and all were treated alike ; the result
was—

TABLE 3.

—	Number of Plants.	Number of Umbels.	Weight of Seed in grains.
Short-styled cowslips . .	47	173	745
Long-styled cowslips . .	58	208	692

These figures give us the following proportions :—

TABLE 4.

—	Number of Plants.	Weight of Seed in grains.	Number of Umbels.	Weight of Seed in grains.
Short-styled cowslips . . .	100	1585	100	430
Long-styled cowslips . . .	100	1093	100	332

The season was much more favourable this year than
the last; the plants also now grew in good soil, instead
of in a shady wood or struggling with other plants in
the open field ; consequently the actual produce of
seed was considerably larger. Nevertheless we have
the same relative result; for the short-styled plants
produced more seed than the long-styled in nearly the
proportion of three to two; but if we take the fairest
standard of comparison, namely, the product of seeds
from an equal number of umbels, the excess is, as in
the former case, nearly as four to three.

Looking to these trials made during two successive
years on a large number of plants, we may safely conclude that the short-styled form is more productive
than the long-styled form, and the same result holds

good with some other species of Primula. Consequently my anticipation that the plants with longer pistils, rougher stigmas, shorter stamens and smaller pollen-grains, would prove to be more feminine in nature, is exactly the reverse of the truth.

In 1860 a few umbels on some plants of both the long-styled and short-styled form, which had been covered by a net, did not produce any seed, though other umbels on the same plants, artificially fertilised, produced an abundance of seed; and this fact shows that the mere covering in itself was not injurious. Accordingly, in 1861, several plants were similarly covered just before they expanded their flowers; these turned out as follows :—

TABLE 5.

—	Number of Plants.	Number of Umbels produced.	Product of Seed.
Short-styled . . .	6	24	1·3 grain weight of seed, or about 50 in number.
Long-styled . . .	18	74	Not one seed.

Judging from the exposed plants which grew all round in the same bed, and had been treated in the same manner, excepting that they had been exposed to the visits of insects, the above six short-styled plants ought to have produced 92 grains' weight of seed instead of only 1·3; and the eighteen long-styled plants, which produced not one seed, ought to have produced above 200 grains' weight. The production of a few seeds by the short-styled plants was probably due to the action of Thrips or of some other minute insect. It is scarcely necessary to give any additional evidence, but I may add that ten pots of polyanthuses and cowslips of both forms, protected from insects in my greenhouse, did not set one pod, though artificially fertilised flowers in other pots produced an abundance. We thus see that the visits of insects are absolutely necessary for the fertilisation of *Primula veris*. If the corolla of the long-styled form had dropped off, instead of remaining attached in a withered state to the ovarium, the anthers attached to the lower part of the tube with some pollen still adhering to them would have been dragged over the stigma, and the flowers would have been partially self-fertilised, as is the case with *Primula Sinensis* through this means. It is a rather curious fact that so trifling a difference as the falling-off of the withered corolla, should make a very great difference in the number of seeds produced by a plant if its flowers are not visited by insects.

The flowers of the cowslip and of the other species of the genus secrete plenty of nectar; and I have often seen humble-bees, especially *B. hortorum* and *muscorum*, sucking the former in a proper manner,* though they sometimes bite holes through the corolla. No doubt moths likewise visit the flowers, as one of my sons caught *Cucullia verbasci* in the act. The pollen readily adheres to any thin object which is inserted into a flower. The anthers in the one form stand nearly, but not exactly, on a level with the stigma of the other; for the distance between the anthers and stigma in the short-styled form is greater than that in the long-styled, in the ratio of 100 to 90. This difference is the result of the anthers in the long-styled form standing rather higher in the tube than does the stigma in the short-styled, and this favours their

---

* H. Müller has also seen *Anthophora pilipes* and a Bombylius sucking the flowers. 'Nature,' Dec. 10th, 1874, p. 111.

pollen being deposited on it. It follows from the position of the organs that if the proboscis of a dead humble-bee, or a thick bristle or rough needle, be pushed down the corolla, first of one form and then of the other, as an insect would do in visiting the two forms growing mingled together, pollen from the long-stamened form adheres round the base of the object, and is left with certainty on the stigma of the long-styled form; whilst pollen from the short stamens of the long-styled form adheres a little way above the extremity of the object, and some is generally left on the stigma of the other form. In accordance with this observation I found that the two kinds of pollen, which could easily be recognised under the microscope, adhered in this manner to the proboscides of the two species of humble-bees and of the moth, which were caught visiting the flowers; but some small grains were mingled with the larger grains round the base of the proboscis, and conversely some large grains with the small grains near the extremity of the proboscis. Thus pollen will be regularly carried from the one form to the other, and they will reciprocally fertilise one another. Nevertheless an insect in withdrawing its proboscis from the corolla of the long-styled form cannot fail occasionally to leave pollen from the same flower on the stigma; and in this case there might be self-fertilisation. But this will be much more likely to occur with the short-styled form; for when I inserted a bristle or other such object into the corolla of this form, and had, therefore, to pass it down between the anthers seated round the mouth of the corolla, some pollen was almost invariably carried down and left on the stigma. Minute insects, such as Thrips, which sometimes haunt the flowers, would likewise be apt to cause the self-fertilisation of both forms.

The several foregoing facts led me to try the effects of the two kinds of pollen on the stigmas of the two forms. Four essentially different unions are possible; namely, the fertilisation of the stigma of the long-styled form by its own-form pollen, and by that of the short-styled; and the stigma of the short-styled form by its own-form pollen, and by that of the long-styled. The fertilisation of either form with pollen from the other form may be conveniently called a *legitimate union*, from reasons hereafter to be made clear; and that of either form with its own-form pollen an *illegitimate union*. I formerly applied the term " heteromorphic " to the legitimate unions, and " homomorphic " to the illegitimate unions; but after discovering the existence of trimorphic plants, in which many more unions are possible, these two terms ceased to be applicable. The illegitimate unions of both forms might have been tried in three ways; for a flower of either form may be fertilised with pollen from the same flower, or with that from another flower on the same plant, or with that from a distinct plant of the same form. But to make my experiments perfectly fair, and to avoid any evil result from self-fertilisation or too close interbreeding, I have invariably employed pollen from a distinct plant of the same form for the illegitimate unions of all the species; and therefore it may be observed that I have used the term " own-form pollen " in speaking of such unions. The several plants in all my experiments were treated in exactly the same manner, and were carefully protected by fine nets from the access of insects, excepting Thrips, which it is impossible to exclude. I performed all the manipulations myself, and weighed the seeds in a chemical balance; but during many subsequent trials I followed the more accurate plan of counting the seeds. Some of the capsules contained no seeds, or only two or three, and these are excluded in the column headed "good capsules" in several of the following tables :—

TABLE 6.

Primula veris.

Nature of the Union.	Number of Flowers fertilised.	Total Number of Capsules produced.	Number of good Capsules.	Weight of Seed in grains.	Calculated Weight of Seed from 100 good Capsules.
Long-styled by pollen of short-styled. Legitimate union . .	22	15	14	8·8	62
Long-styled by own-form pollen. Illegitimate union . .	20	8	5	2·1	42
Short-styled by pollen of long-styled. Legitimate union . .	13	12	11	4·9	44
Short-styled by own-form pollen. Illegitimate union. .	15	8	6	1·8	30
SUMMARY: The two legitimate unions	35	27	25	13·7	54
The two illegitimate unions	35	16	11	3·9	35

The results may be given in another form (Table 7) by comparing, first, the number of capsules, whether good or bad, or of the good alone, produced by 100 flowers of both forms when legitimately and illegitimately fertilised; secondly, by comparing the weight of seed in 100 of these capsules, whether good or bad; or, thirdly, in 100 of the good capsules.

TABLE 7.

Nature of the Union.	Number of Flowers fertilised.	Number of Capsules.	Number of good Capsules.	Weight of Seed in grains.	Number of Capsules.	Weight of Seed in grains.	Number of good Capsules.	Weight of Seed in grains.
The two legitimate unions .	100	77	71	39	100	50	100	54
The two illegitimate unions .	100	45	31	11	100	24	100	35

We here see that the long-styled flowers fertilised
with pollen from the short-styled yield more capsules,
especially good ones (i.e. containing more than one
or two seeds), and that these capsules contain a greater
proportional weight of seeds than do the flowers of the
long-styled when fertilised with pollen from a distinct
plant of the same form. So it is with the short-styled
flowers, if treated in an analogous manner. Therefore I
have called the former method of fertilisation a legitimate union, and the latter, as it fails to yield the full
complement of capsules and seeds, an illegitimate
union. These two kinds of union are graphically represented in Fig. 2.

If we consider the results of the two legitimate
unions taken together and the two illegitimate ones,
as shown in Table 7, we see that the former compared with the latter yielded capsules, whether containing many seeds or only a few, in the proportion of
77 to 45, or as 100 to 58. But the inferiority of the
illegitimate unions is here perhaps too great, for on a
subsequent occasion 100 long-styled and short-styled
flowers were illegitimately fertilised, and they together
yielded 53 capsules: therefore the rate of 77 to 53, or
as 100 to 69, is a fairer one than that of 100 to 58.

Returning to Table 7, if we consider only the good capsules, those from the two legitimate unions were to those from the two illegitimate in number as 71 to 31, or as 100 to 44. Again, if we take an equal number of capsules, whether good or bad, from the legitimately and illegitimately fertilised flowers, we find that the former contained seeds by weight compared with the latter as 50 to 24, or as 100 to 48; but if all the

Fig. 2.

Legitimate union.
Complete fertility.

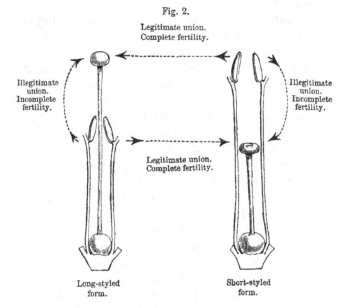

Illegitimate
union.
Incomplete
fertility.

Illegitimate
union.
Incomplete
fertility.

Legitimate union.
Complete fertility.

Long-styled
form.

Short-styled
form.

poor capsules are rejected, of which many were produced by the illegitimately fertilised flowers, the proportion is 54 to 35, or as 100 to 65. In this and all other cases, the relative fertility of the two kinds of union can, I think, be judged of more truly by the average number of seeds per capsule than by the proportion of flowers which yield capsules. The two methods might

have been combined by giving the average number of
seeds produced by all the flowers which were fertilised,
whether they yielded capsules or not; but I have
thought that it would be more instructive always to
show separately the proportion of flowers which produced capsules, and the average number of apparently
good seeds which the capsules contained.

Flowers legitimately fertilised set seeds under conditions which cause the almost complete failure of
illegitimately fertilised flowers. Thus in the spring of
1862 forty flowers were fertilised at the same time in
both ways. The plants were accidentally exposed in
the greenhouse to too hot a sun, and a large number
of umbels perished. Some, however, remained in moderately good health, and on these there were twelve
flowers which had been fertilised legitimately, and
eleven which had been fertilised illegitimately. The
twelve legitimate unions yielded seven fine capsules,
containing on an average each 57·3 good seeds; whilst
the eleven illegitimate unions yielded only two capsules, of which one contained 39 seeds, but so poor,
that I do not suppose one would have germinated, and
the other contained 17 fairly good seeds.

From the facts now given the superiority of a legitimate over an illegitimate union admits of not the
least doubt; and we have here a case to which no
parallel exists in the vegetable or, indeed, in the
animal kingdom. The individual plants of the present species, and as we shall see of several other
species of Primula, are divided into two sets or
bodies, which cannot be called distinct sexes, for
both are hermaphrodites; yet they are to a certain
extent sexually distinct, for they require reciprocal
union for perfect fertility. As quadrupeds are divided into two nearly equal bodies of different sexes,

so here we have two bodies, approximately equal in number, differing in their sexual powers and related to each other like males and females. There are many hermaphrodite animals which cannot fertilise themselves, but must unite with another hermaphrodite. So it is with numerous plants; for the pollen is often mature and shed, or is mechanically protruded, before the flower's own stigma is ready; and such flowers absolutely require the presence of another hermaphrodite for sexual union. But with the cowslip and various other species of Primula there is this wide difference, that one individual, though it can fertilise itself imperfectly, must unite with another individual for full fertility; it cannot, however, unite with any other individual in the same manner as an hermaphrodite plant can unite with any other one of the same species ; or as one snail or earth-worm can unite with any other hermaphrodite individual. On the contrary, an individual belonging to one form of the cowslip in order to be perfectly fertile must unite with one of the other form, just as a male quadruped must and can unite only with the female.

I have spoken of the legitimate unions as being fully fertile; and I am fully justified in doing so, for flowers artificially fertilised in this manner yielded rather more seeds than plants naturally fertilised in a state of nature. The excess may be attributed to the plants having been grown separately in good soil. With respect to the illegitimate unions, we shall best appreciate their degree of lessened fertility by the following facts. Gärtner estimated the sterility of the unions between distinct species,* in a manner which allows of a strict comparison with the results of the

---

* 'Versuche über die Bastarderzeugung,' 1849, p. 216.

legitimate and illegitimate unions of Primula. With
P. veris, for every 100 seeds yielded by the two legitimate unions, only 64 were yielded by an equal
number of good capsules from the two illegitimate
unions. With *P. Sinensis*, as we shall hereafter see,
the proportion was nearly the same—namely, as 100
to 62. Now Gärtner has shown that, on the calculation of *Verbascum lychnitis* yielding with its own pollen
100 seeds, it yielded when fertilised by the pollen of
V. Phœniceum 90 seeds; by the pollen of *V. nigrum*,
63 seeds; by that of *V. blattaria*, 62 seeds. So again,
Dianthus barbatus fertilised by the pollen of *D. superbus*
yielded 81 seeds, and by the pollen of *D. Japonicus*
66 seeds, relatively to the 100 seeds produced by its
own pollen. We thus see—and the fact is highly remarkable—that with Primula the illegitimate unions
relatively to the legitimate are more sterile than
crosses between distinct species of other genera relatively to their pure unions. Mr. Scott has given[*] a
still more striking illustration of the same fact: he
crossed *Primula auricula* with pollen of four other
species (*P. Palinuri, viscosa, hirsuta*, and *verticillata*),
and these hybrid unions yielded a larger average
number of seeds than did *P. auricula* when fertilised
illegitimately with its own-form pollen.

The benefit which heterostyled dimorphic plants de-
rive from the existence of the two forms is sufficiently
obvious, namely, the intercrossing of distinct plants
being thus ensured.[†] Nothing can be better adapted
for this end than the relative positions of the anthers
and stigmas in the two forms, as shown in Fig. 2; but to

[*] 'Journ. Linn. Soc. Bot.,' vol. viii. 1864, p. 93.

[†] I have shown in my work on the 'Effects of Cross and Self- fertilisation' how greatly the off- spring from intercrossed plants profit in height, vigour, and fertility.

this whole subject I shall recur. No doubt pollen will
occasionally be placed by insects or fall on the stigma
of the same flower; and if cross-fertilisation fails, such
self-fertilisation will be advantageous to the plant, as
it will thus be saved from complete barrenness. But
the advantage is not so great as might at first be
thought, for the seedlings from illegitimate unions do
not generally consist of both forms, but all belong to
the parent form; they are, moreover, in some degree
weakly in constitution, as will be shown in a future
chapter. If, however, a flower's own pollen should first
be placed by insects or fall on the stigma, it by no
means follows that cross-fertilisation will be thus prevented. It is well known that if pollen from a distinct
species be placed on the stigma of a plant, and some
hours afterwards its own pollen be placed on it, the
latter will be prepotent and will quite obliterate any
effect from the foreign pollen; and there can hardly
be a doubt that with heterostyled dimorphic plants,
pollen from the other form will obliterate the effects of
pollen from the same form, even when this has been
placed on the stigma a considerable time before. To
test this belief, I placed on several stigmas of a long-
styled cowslip plenty of pollen from the same plant,
and after twenty-four hours added some from a short-
styled dark-red polyanthus, which is a variety of the
cowslip. From the flowers thus treated 30 seedlings
were raised, and all these, without exception, bore
reddish flowers; so that the effect of pollen from the
same form, though placed on the stigmas twenty-four
hours previously, was quite destroyed by that of pollen
from a plant belonging to the other form.

Finally, I may remark that of the four kinds of
unions, that of the short-styled illegitimately fertilised
with its own-form pollen seems to be the most sterile of

all, as judged by the average number of seeds, which the capsules contained. A smaller proportion, also, of these seeds than of the others germinated, and they germinated more slowly. The sterility of this union is the more remarkable, as it has already been shown that the short-styled plants yield a larger number of seeds than the long-styled, when both forms are fertilised, either naturally or artificially, in a legitimate manner.

In a future chapter, when I treat of the offspring from heterostyled dimorphic and trimorphic plants illegitimately fertilised with their own-form pollen, I shall have occasion to show that with the present species and several others, equal-styled varieties sometimes appear.

Primula elatior, Jacq.

Bardfield Oxlip of English Authors.

This plant, as well as the last or Cowslip (*P. veris,* vel *officinalis*), and the Primrose (*P.vulgaris,* vel *acaulis*) have been considered by some botanists as varieties of the same species. But they are all three undoubtedly distinct, as will be shown in the next chapter. The present species resembles to a certain extent in general appearance the common oxlip, which is a hybrid between the cowslip and primrose. *Primula elatior* is found in England only in two or three of the eastern counties; and I was supplied with living plants by Mr. Doubleday, who, as I believe, first called attention to its existence in England. It is common in some parts of the Continent; and H. Müller* has seen several kinds of humble-bees and other bees, and Bombylius, visiting the flowers in North Germany.

---

* 'Die Befruchtung der Blumen,' p. 347.

The results of my trials on the relative fertility of the two forms, when legitimately and illegitimately fertilised, are given in the following table :—

TABLE 8.

Primula elatior.

Nature of Union.	Number of Flowers fertilised.	Number of good Capsules produced.	Maximum of Seeds in any one Capsule.	Minimum of Seeds in any one Capsule.	Average Number of Seeds per Capsule.
Long-styled form, by pollen of short-styled. Legitimate union .	10	6	62	34	46·5
Long-styled form, by own-form pollen. Illegitimate union . .	20	4	49*	2	27·7
Short-styled form, by pollen of long-styled. Legitimate union .	10	8	61	37	47·7
Short-styled form, by own-form pollen. Illegitimate union. .	17	3	19	9	12·1
The two legitimate unions together . .	20	14	62	37	47·1
The two illegitimate unions together . .	37	7	49*	2	35·5

* These seeds were so poor and small that they could hardly have germinated.

If we compare the fertility of the two legitimate unions taken together with that of the two illegitimate unions together, as judged by the proportional number of flowers which when fertilised in the two methods yielded capsules, the ratio is as 100 to 27; so that by this standard the present species is much more sterile than *P. veris*, when both species are illegitimately fertilised. If we judge of the relative fertility of the two kinds of unions by the average number of seeds per capsule, the ratio is as 100 to 75. But this latter

D

number is probably much too high, as many of the seeds
produced by the illegitimately fertilised long-styled
flowers were so small that they probably would not
have germinated, and ought not to have been counted.
Several long-styled and short-styled plants were protected from the access of insects, and must have been
spontaneously self-fertilised. They yielded altogether
only six capsules, containing any seeds; and their
average number was only 7·8 per capsule. Some,
moreover, of these seeds were so small that they could
hardly have germinated.

Herr W. Breitenbach informs me that he examined,
in two sites near the Lippe (a tributary of the Rhine),
894 flowers produced by 198 plants of this species; and
he found 467 of these flowers to be long-styled, 411
short-styled, and 16 equal-styled. I have heard of no
other instance with heterostyled plants of equal-styled
flowers appearing in a state of nature, though far from
rare with plants which have been long cultivated. It
is still more remarkable that in eighteen cases the
same plant produced both long-styled and short-styled,
or long-styled and equal-styled flowers; and in two
out of the eighteen cases, long-styled, short-styled, and
equal-styled flowers. The long-styled flowers greatly
preponderated on these eighteen plants,—61 consisting
of this form, 15 of equal-styled, and 9 of the short-
styled form.

PRIMULA VULGARIS (var. *acaulis*, Linn.),

The Primrose of English Writers.

Mr. J. Scott examined 100 plants growing near
Edinburgh, and found 44 to be long-styled, and 56
short-styled; and I took by chance 79 plants in Kent,
of which 39 were long-styled and 40 short-styled; so

that the two lots together consisted of 83 long-styled
and 96 short-styled plants. In the long-styled form
the pistil is to that of the short-styled in length, from
an average of five measurements, as 100 to 51. The
stigma in the long-styled form is conspicuously more
globose and much more papillose than in the short-
styled, in which latter it is depressed on the summit;

Fig. 3.

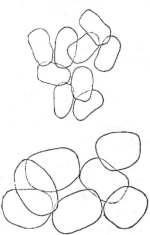

Outlines of pollen-grains of *Primula vulgaris*, distended with water, much
magnified and drawn under the camera lucida. The upper and smaller
grains from the long-styled form; the lower and larger grains from
the short-styled.

it is equally broad in the two forms. In both it stands
nearly, but not exactly, on a level with the anthers of
the opposite form; for it was found, from an average
of 15 measurements, that the distance between the
middle of the stigma and the middle of the anthers
in the short-styled form is to that in the long-styled
as 100 to 93. The anthers do not differ in size in the
two forms. The pollen-grains from the short-styled

flowers before they were soaked in water were decidedly broader, in proportion to their length, than those from the long-styled; after being soaked they were relatively to those from the long-styled as 100 to 71 in diameter, and more transparent. A large number of flowers from the two forms were compared, and 12 of the finest flowers from each lot were measured, but there was no sensible difference between them in size. Nine long-styled and eight short-styled plants growing together in a state of nature were marked, and their capsules collected after they had been naturally fertilised; and the seeds from the short-styled weighed exactly twice as much as those from an equal number of long-styled plants. So that the primrose resembles the cowslip in the short-styled plants, being the more productive of the two forms. The results of my trials on the fertility of the two forms, when legitimately and illegitimately fertilised, are given in Table 9.

We may infer from this table that the fertility of the two legitimate unions taken together is to that of the two illegitimate unions together, as judged by the proportional number of flowers which when fertilised in the two methods yielded capsules, as 100 to 60. If we judge by the average number of seeds per capsule produced by the two kinds of unions, the ratio is as 100 to 54; but this latter figure is perhaps rather too low. It is surprising how rarely insects can be seen during the day visiting the flowers, but I have occasionally observed small kinds of bees at work; I suppose, therefore, that they are commonly fertilised by nocturnal Lepidoptera. The long-styled plants when protected from insects yield a considerable number of capsules, and they thus differ remarkably from the same form of the cowslip, which is quite sterile under the same circumstances. Twenty-three spontaneously self-fertilised capsules from

TABLE 9.

Primula vulgaris.

Nature of Union.	Number of Flowers fertilised.	Number of good Capsules produced.	Maximum Number of Seeds in any one Capsule.	Minimum Number of Seeds in any one Capsule.	Average Number of Seeds per Capsule.
Long-styled form, by pollen from short-styled. Legitimate union . .	12	11	77	47	66·9
Long-styled form, by own-form pollen. Illegitimate union . .	21	14	66	30	52·2
Short-styled form, by pollen from long-styled. Legitimate union . .	8	7	75	48	65·0
Short-styled form, by own-form pollen. Illegitimate union . .	18	7	43	5	18·8*
The two legitimate unions together . .	20	18	77	47	66·0
The two illegitimate unions together . .	39	21	66	5	35·5*

* This average is perhaps rather too low.

this form contained, on an average, 19·2 seeds. The short-styled plants produced fewer spontaneously self-fertilised capsules, and fourteen of them contained only 6·2 seeds per capsule. The self-fertilisation of both forms was probably aided by Thrips, which abounded within the flowers; but these minute insects could not have placed nearly sufficient pollen on the stigmas, as the spontaneously self-fertilised capsules contained much fewer seeds, on an average, than those (as may be seen in Table 9) which were artificially fertilised with their own-form pollen. But this difference may perhaps be attributed in part to the flowers in the table having been fertilised with pollen from a distinct plant belonging to the same form; whilst those which were spontaneously self-fertilised no doubt generally received their own pollen. In a future part of this volume some observations will be given on the fertility of a red-coloured variety of the primrose.

PRIMULA SINENSIS.

In the long-styled form the pistil is about twice as long as that of the short-styled, and the stamens differ in a corresponding, but reversed, manner. The stigma is considerably more elongated and rougher than that of the short-styled, which is smooth and almost spherical, being somewhat depressed on the summit; but the stigma varies much in all its characters, the result, probably, of cultivation. The pollen-grains of the short-styled form, according to Hildebrand,[*] are 7 divisions of the micrometer in length and 5 in breadth; whereas those of the long-styled are only 4 in length and 3 in breadth. The grains, therefore, of the short-styled are to those of the long-styled in length as 100 to 57. Hildebrand also remarked, as I had done in the case of *P. veris*, that the smaller grains from the long-styled are much more transparent than the larger ones from the short-styled form. We shall hereafter see that this cultivated plant varies much in its dimorphic condition and is often equal-styled. Some individuals may be said to be sub-heterostyled; thus in two white-flowered plants the pistil projected above the stamens, but in one of them

---

[*] After the appearance of my paper this author published some excellent observations on the present species ('Bot. Zeitung,' Jan. 1, 1864), and he shows that I erred greatly about the size of the pollen-grains in the two forms. I suppose that by mistake I measured twice over pollen-grains from the same form

it was longer and had a more elongated and rougher
stigma, than in the other; and the pollen-grains from
the latter were to those from the plant with a more
elongated pistil only as 100 to 88 in diameter, instead
of as 100 to 57. The corolla of the long-styled and
short-styled form differs in shape, in the same manner
as in *P. veris*. The long-styled plants tend to flower
before the short-styled. When both forms were legitimately fertilised, the capsules from the short-styled
plants contained, on an average, more seeds than those
from the long-styled, in the ratio of 12·2 to 9·3 by
weight, that is, as 100 to 78. In the following table
we have the results of two sets of experiments tried
at different periods.

TABLE 10.

Primula Sinensis.

Nature of Union.	Number of Flowers fertilised.	Number of good Capsules produced.	Average Weight of Seeds per Capsule.	Average Number of Seeds per Capsule, as ascertained on a subsequent occasion.
Long-styled form, by pollen of short-styled. Legitimate union	24	16	0·58	50
Long-styled form, by own-form pollen. Illegitimate union	20	13	0·45	35
Short-styled form, by pollen of long-styled. Legitimate union . . .	8	8	0·76	64
Short-styled form, by own-form pollen. Illegitimate union	7	4	0·23	25
The two legitimate unions together	32	24	0·64	57
The two illegitimate unions together	27	17	0·40	30

The fertility, therefore, of the two legitimate unions together to that of the two illegitimate unions, as judged by the proportional number of flowers which yielded capsules, is as 100 to 84. Judging by the average weight of seeds per capsule produced by the two kinds of unions, the ratio is as 100 to 63. On another occasion a large number of flowers of both forms were fertilised in the same manner, but no account of their number was kept. The seeds, however, were carefully counted, and the averages are shown in the right-hand column. The ratio for the number of seeds produced by the two legitimate compared with the two illegitimate unions is here 100 to 53, which is probably more accurate than the foregoing one of 100 to 63.

Hildebrand in the paper above referred to gives the results of his experiments on the present species; and these are shown in a condensed form in the following table (11). Besides using for the illegitimate unions pollen from a distinct plant of the same form, as was always done by me, he tried, in addition, the effects of the plant's own pollen. He counted the seeds.

It is remarkable that here all the flowers which were fertilised legitimately, as well as those fertilised illegitimately with pollen from a distinct plant belonging to the same form, yielded capsules; and from this fact it might be inferred that the two forms were reciprocally much more fertile in his case than in mine. But his illegitimately fertilised capsules from both forms contained fewer seeds relatively to the legitimately fertilised capsules than in my experiments; for the ratio in his case is as 42 to 100, instead of, as in mine, as 53 to 100. Fertility is a very variable element with most plants, being determined by the conditions to which they are subjected, of which fact I have observed striking instances with the

TABLE 11.

Primula Sinensis (from Hildebrand).

Nature of Union.	Number of Flowers fertilised.	Number of good Capsules produced.	Average Number of Seeds per Capsule.
Long-styled form, by pollen of short-styled. Legitimate union	14	14	41
Long-styled form, by own-form pollen, from a distinct plant. Illegitimate union .	26	26	18
Long-styled form, by pollen from same flower. Illegitimate union. . . .	27	21	17
Short-styled form, by pollen of long-styled. Legitimate union	14	14	44
Short-styled form, by own-form pollen, from a distinct plant. Illegitimate union	16	16	20
Short-styled, by pollen from the same flower. Illegitimate union. . . .	21	11	8
The two legitimate unions together . .	28	28	43
The two illegitimate unions together (own-form pollen)	42	42	18
The two illegitimate unions together (pollen from the same flower)	48	32	13

present species; and this may account for the difference between my results and those of Hildebrand. His plants were kept in a room, and perhaps were grown in too small pots or under some other unfavourable conditions, for his capsules in almost every case contained a smaller number of seeds than mine, as may be seen by comparing the right-hand columns in Tables 10 and 11.

The most interesting point in Hildebrand's experiments is the difference in the effects of illegitimate fertilisation with a flower's own pollen, and with that

from a distinct plant of the same form.   In the latter
case all the flowers produced capsules, whilst only 67
out of 100 of those fertilised with their own pollen produced capsules.   The self-fertilised capsules also contained seeds, as compared with capsules from flowers
fertilised with pollen from a distinct plant of the same
form, in the ratio of 72 to 100.

In order to ascertain how far the present species was
spontaneously self-fertile, five long-styled plants were
protected by me from insects; and they bore up to a
given period 147 flowers which set 62 capsules; but
many of these soon fell off, showing that they had not
been properly fertilised.   At the same time five short-
styled plants were similarly treated, and they bore 116
flowers which ultimately produced only seven capsules.
On another occasion 13 protected long-styled plants
yielded by weight 25·9 grains of spontaneously self-
fertilised seeds.   At the same time seven protected
short-styled plants yielded only half-a-grain weight of
seeds.   Therefore the long-styled plants yielded nearly
24 times as many spontaneously self-fertilised seeds as
did the same number of short-styled plants.   The chief
cause of this great difference appears to be, that when
the corolla of a long-styled plant falls off, the anthers,
from being situated near the bottom of the tube are
necessarily dragged over the stigma and leave pollen
on it, as I saw when I hastened the fall of nearly
withered flowers; whereas in the short-styled flowers,
the stamens are seated at the mouth of the corolla,
and in falling off do not brush over the lowly-seated
stigmas.   Hildebrand likewise protected some long-
styled and short-styled plants, but neither ever yielded
a single capsule.   He thinks that the difference in our
results may be accounted for by his plants having
been kept in a room and never having been shaken;

but this explanation seems to me doubtful ; his plants were in a less fertile condition than mine, as shown by the difference in the number of seeds produced, and it is highly probable that their lessened fertility would have interfered with especial force with their capacity for producing self-fertilised seeds.

PRIMULA AURICULA.[*]

This species is heterostyled, like the preceding ones; but amongst the varieties distributed by florists the long-styled form is rare, as it is not valued. There is a much greater relative inequality in the length of the pistil and stamens in the two forms of the auricula than in the cowslip; the pistil in the long-styled being nearly four times as long as that in the short-styled, in which it is barely longer than the ovarium. The stigma is nearly of the same shape in both forms, but is rougher in the long-styled, though the difference is not so great as between the two forms of the cowslip. In the long-styled plants the stamens are very short, rising but little above the ovarium. The pollen-grains of these short stamens, when distended with water, were barely $\frac{5}{6000}$ of an inch in diameter, whereas those from the long stamens of the short-styled plants were barely $\frac{7}{6000}$, showing a relative difference of about 71 to 100. The smaller grains of the long-styled plant are also much more transparent, and before distention with water more triangular in outline than those of the other form. Mr. Scott [†] compared ten plants of both forms growing under similar conditions, and found that, although the long-styled plant produced more umbels and more capsules than the short-styled, yet they yielded fewer seeds, in the ratio of 66 to 100. Three short-styled plants were protected by me from the

---

* According to Kerner our garden auriculas are descended from *P. pubescens*, Jacq., which is a hybrid between the true *P. auricula* and *hirsuta*. This hybrid has now been propagated for about 300 years, and produces, when legitimately fertilised, a large number of seeds; the long-styled forms yielding an average number of 73, and the short-styled 98 seeds per capsule: see his "Geschichte der Aurikel," ' Zeitschr. des Deutschen und Oest. Alpen-Vereins,' Band vi. p. 52. Also ' Die Primulaceen-Bastarten,' ' Oest. Bot. Zeitschrift,' 1835, Nos. 3, 4, and 5.

† ' Journ. Linn. Soc. Bot.' vol. viii. 1864, p. 86.

access of insects, and they did not produce a single seed. Mr. Scott protected six plants of both forms, and found them excessively sterile. The pistil of the long-styled form stands so high above the anthers, that it is scarcely possible that pollen should reach the stigma without some aid; and one of Mr. Scott's long-styled plants which yielded a few seeds (only 18 in number) was infested by aphides, and he does not doubt that these had imperfectly fertilised it.

I tried a few experiments by reciprocally fertilising the two forms in the same manner as before, but my plants were unhealthy, so I will give, in a condensed form, the results of Mr. Scott's experiments. For fuller particulars with respect to this and the five following species, the paper lately referred to may be consulted. In each case the fertility of the two legitimate unions, taken together, is compared with that of the two illegitimate unions together, by the same two standards as before, namely, by the proportional number of flowers which produced good capsules, and by the average number of seeds per capsule. The fertility of the legitimate unions is always taken at 100.

By the first standard, the fertility of the two legitimate unions of the auricula is to that of the two illegitimate unions as 100 to 80; and by the second standard as 100 to 15.

Primula Sikkimensis.

According to Mr. Scott, the pistil of the long-styled form is fully four times as long as that of the short-styled, but their stigmas are nearly alike in shape and roughness. The stamens do not differ so much in relative length as the pistils. The pollen-grains differ in a marked manner in the two forms; "those of the long-styled plants are sharply triquetrous, smaller, and more transparent than those of the short-styled, which are of a bluntly triangular form." The fertility of the two legitimate unions to that of the two illegitimate unions is by the first standard as 100 to 95, and by the second standard as 100 to 31.

Primula Cortusoides.

The pistil of the long-styled form is about thrice as long as that of the short-styled, the stigma being double as long and covered with much longer papillæ. The pollen-grains of the short-styled form are, as usual, "larger, less transparent, and more bluntly triangular than those from the long-styled plants." The fertility of the two legitimate unions to that of the two illegitimate unions is by the first standard as 100 to 74, and by the second standard as 100 to 66.

PRIMULA INVOLUCRATA.

The pistil of the long-styled form is about thrice as long as that of the short-styled; the stigma of the former is globular and closely beset with papillæ, whilst that of the short-styled is smooth and depressed on the apex. The pollen-grains of the two forms differ in size and transparency as before, but not in shape. The fertility of the two legitimate to that of the two illegitimate unions is by the first standard as 100 to 72; and by the second standard as 100 to 47.

PRIMULA FARINOSA.

According to Mr. Scott, the pistil of the long-styled form is only about twice as long as that of the short-styled. The stigmas of the two forms differ but little in shape. The pollen-grains differ in the usual manner in size, but not in form. The fertility of the two legitimate to that of the two illegitimate unions is by the first standard as 100 to 71, and by the second standard as 100 to 44.

Summary on the foregoing heterostyled species of Primula.—The fertility of the long and short-styled plants of the above species of Primula, when the two forms are fertilised legitimately, and illegitimately with pollen of the same form taken from a distinct plant, has now been given. The results are seen in the following table; the fertility being judged by two standards, namely, by that of the proportional number of flowers which yielded capsules, and by that of the average number of seeds per capsule. But for full accuracy many more observations, under varied conditions, would be requisite.

TABLE 12.

Summary on the Fertility of the two Legitimate Unions, compared with that of the two Illegitimate Unions, in the genus Primula. The former taken at 100.

Name of Species.	Illegitimate Unions.	
	Judged of by the Proportional Number of Flowers which produced Capsules.	Judged of by the Average Number (or Weight in some cases) of Seeds per Capsule.
Primula veris	69	65
P. elatior	27	75 {(Probably too high.)
P. vulgaris	60	54 {(Perhaps too low.)
P. Sinensis	84	63
„ (second trial) . .	?	53
„ (after Hildebrand) .	100	42
P. auricula (Scott)	80	15
P. Sikkimensis (Scott) . . .	95	31
P. cortusoides (Scott) . . .	74	66
P. involucrata (Scott) . . .	72	48
P. farinosa (Scott)	71	44
Average of the nine species .	88·4	61·8

With plants of all kinds some flowers generally fail to produce capsules, from various accidental causes; but this source of error has been eliminated, as far as possible, in all the previous cases, by the manner in which the calculations have been made. Supposing, for instance, that 20 flowers were fertilised legitimately and yielded 18 capsules, and that 30 flowers were fertilised illegitimately and yielded 15 capsules, we may assume that on an average an equal proportion of the flowers in both lots would fail to produce capsules from various accidental causes; and the ratio of $\frac{18}{20}$ to $\frac{15}{30}$, or as 100 to 56 (in whole numbers), would show the proportional number of capsules due to the two methods of fertilisation; and the number 56 would appear in the left-hand column of Table 12, and in my other tables. With respect to the average number of seeds per capsule hardly anything need be said : supposing that the legitimately fertilised capsules contained, on an average, 50 seeds, and the illegitimately fertilised capsules 25 seeds; then as 50 is to 25 so is 100 to 50; and the latter number would appear in the right-hand column.

It is impossible to look at the above table and doubt that the legitimate unions between the two forms of the above nine species of Primula are much more fertile than the illegitimate unions; although in the latter case pollen was always taken from a distinct plant of the same form. There is, however, no close correspondence in the two rows of figures, which give, according to the two standards, the difference of fertility between the legitimate and illegitimate unions. Thus all the flowers of *P. Sinensis* which were illegitimately fertilised by Hildebrand produced capsules; but these contained only 42 per cent. of the number of seeds yielded by the legitimately fertilised capsules. So again, 95 per cent. of the illegitimately fertilised flowers of *P. Sikkimensis* produced capsules; but these contained only 31 per cent. of the number of seeds in the legitimate capsules. On the other hand, with *P. elatior* only 27 per cent. of the illegitimately fertilised flowers yielded capsules; but these contained nearly 75 per cent. of the legitimate number of seeds. It appears that the setting of the flowers, that is, the production of capsules whether good or bad, is not so much influenced by legitimate and illegitimate fertilisation as is the number of seeds which the capsules

contain. For, as may be seen at the bottom of Table
12, 88·4 per cent. of the illegitimately fertilised
flowers yielded capsules; but these contained only
61·8 per cent. of seeds, in comparison, in each case,
with the legitimately fertilised flowers and capsules
of the same species.

There is another point which deserves notice,
namely, the relative degree of infertility in the several
species of the long-styled and short-styled flowers,
when both are illegitimately fertilised. The data
may be found in the earlier tables, and in those given
by Mr. Scott in the Paper already referred to. If we
call the number of seeds per capsule produced by the
illegitimately fertilised long-styled flowers 100, the
seeds from the illegitimately fertilised short-styled
flowers will be represented by the following numbers :—

Primula veris.	71		Primula auricula.	. .	119
P. elatior .	. 44	{(Probably too low.)	P. Sikkimensis	. . .	57
			P. cortusoides	. . .	93
P. vulgaris	. 36	{(Perhaps too low.)	P. involucrata	. . .	74
P. Sinensis	. 71		P. farinosa	. . . .	63

We thus see that, with the exception of *P. auricula*, the
long-styled flowers of all nine species are more fertile
than the short-styled flowers, when both forms are
illegitimately fertilised. Whether *P. auricula* really
differs from the other species in this respect I can form
no opinion, as the result may have been accidental.
The degree of self-fertility of a plant depends on two
elements, namely, on the stigma receiving its own pollen
and on its more or less efficient action when placed
there. Now as the anthers of the short-styled flowers of
several species of Primula stand directly above the
stigma, their pollen is more likely to fall on it, or to
be carried down to it by insects, than in the case of

the long-styled form. It appears probable, therefore,
at first sight, that the lessened capacity of the short-
styled flowers to be fertilised with their own pollen, is
a special adaptation for counteracting their greater
liability to receive their own pollen, and thus for
checking self-fertilisation. But from facts with respect
to other species hereafter to be given, this view can
hardly be admitted. In accordance with the above
liability, when some of the species of Primula were
allowed to fertilise themselves spontaneously under
a net, all insects being excluded, except such minute
ones as Thrips, the short-styled flowers, notwithstanding their greater innate self-sterility, yielded
more seed than did the long-styled. None of the
species, however, when insects were excluded, made
a near approach to full fertility. But the long-styled
form of *P. Sinensis* gave, under these circumstances,
a considerable number of seeds, as the corolla in falling
off drags the anthers, which are seated low down in
the tube, over the stigma, and thus leaves plenty of
pollen on it.

Homostyled species of Primula.—It has now been
shown that nine of the species in this genus exist under
two forms, which differ not only in structure but in
function. Besides these Mr. Scott enumerates 27 other
species* which are heterostyled ; and to these probably
others will be hereafter added. Nevertheless, some
species are homostyled ; that is, they exist only under
a single form ; but much caution is necessary on this
head, as several species when cultivated are apt to
become equal-styled. Mr. Scott believes that *P.
Scotica, verticillata,* a variety of *Sibirica, elata, mollis,* and

---

* H. Müller has given in ' Nature,' Dec. 10, 1874, p. 110, a
drawing of one of these species,
viz. the Alpine *P. villosa,* and
shows that it is fertilised exclusively by Lepidoptera.

E

longiflora,* are truly homostyled; and to these may be added, according to Axell, *P. stricta.* Mr. Scott experimented on *P. Scotica, mollis,* and *verticillata,* and found that their flowers yielded an abundance of seeds when fertilised with their own pollen. This shows that they are not heterostyled in function. *P. Scotica* is, however, only moderately fertile when insects are excluded, but this depends merely on the coherent pollen not readily falling on the stigma without their aid. Mr. Scott also found that the capsules of *P. verticillata* contained rather more seed when the flowers were fertilised with pollen from a distinct plant than when with their own pollen; and from this fact he infers that they are sub-heterostyled in function, though not in structure. But there is no evidence that two sets of individuals exist, which differ slightly in function and are adapted for reciprocal fertilisation; and this is the essence of heterostylism. The mere fact of a plant being more fertile with pollen from a distinct individual than with its own pollen, is common to very many species, as I have shown in my work 'On the Effects of Cross and Self-fertilisation.'

HOTTONIA PALUSTRIS.

This aquatic member of the Primulaceæ is conspicuously heterostyled, as the pistil of the long-styled form projects far out of the flower, the stamens being enclosed within the tube; whilst the stamens of the short-styled flower project far outwards, the pistil being enclosed. This difference between the two forms has attracted the attention of various botanists, and that

---

* Koch was aware that this species was homostyled: see " Treviranus über Dichogamie nach Sprengel und Darwin," ' Bot. Zeitung,' Jan. 2, 1863, p. 4.

of Sprengel,* in 1793, who, with his usual sagacity, adds that he does not believe the existence of the two forms to be accidental, though he cannot explain their purpose. The pistil of the long-styled form is more than twice as long as that of the short-styled, with the stigma rather smaller, though rougher. H. Müller† gives figures of the stigmatic papillæ of the two forms, and those of the long-styled are seen to be more than double the length, and much thicker than the papillæ of the short-styled form. The anthers in the one form do not stand exactly on a level with the stigma in the other form; for the distance between the organs is greater in the short-styled than in the long-styled flowers in the proportion of 100 to 71. In dried specimens soaked in water the anthers of the short-styled form are larger than those of the long-styled, in the ratio of 100 to 83. The pollen-grains, also, from the short-styled flowers are conspicuously larger than those from the long-styled; the ratio between the diameters of the moistened grains being as 100 to 64, according to my measurements, but according to the measurements of H. Müller as 100 to 61; and his are probably the more accurate of the two. The contents of the larger pollen-grains appear more coarsely granular and of a browner tint, than those in the smaller grains. The two forms of Hottonia thus agree closely in most respects with those of the heterostyled species of Primula. The flowers of Hottonia are cross-fertilised, according to Müller, chiefly by Diptera.

Mr. Scott‡ made a few trials on a short-styled plant, and found that the legitimate unions were in all ways more fertile than the illegitimate; but since the pub-

---

* 'Das entdeckte Geheimniss der Nature,' p. 108.
† 'Die Befruchtung,' &c., p. 350.
‡ 'Journ. Linn. Soc. Bot.' vol. viii. 1864, p. 79.

lication of his paper H. Müller has made much fuller
experiments, and I give his results in the following
table, drawn up in accordance with my usual plan :—

TABLE 13.

Hottonia palustris (from H. Müller).

Nature of Union.	Number of Capsules examined.	Average Number of Seeds per Capsule.
Long-styled form, by pollen of short-styled. Legitimate union	34	91·4
Long-styled form, by own-form pollen, from a distinct plant. Illegitimate union . . .	18	77·5
Short-styled form, by pollen of long-styled. Legitimate union	30	66·2
Short-styled form, by own-form pollen, from a distinct plant. Illegitimate union . . .	19	18·7
The two legitimate unions together . . .	64	78·8
The two illegitimate unions together . . .	37	48·1

The most remarkable point in this table is the
small average number of seeds from the short-styled
flowers when illegitimately fertilised, and the unusually
large average number of seeds yielded by the illegitimately fertilised long-styled flowers, relatively in both
cases to the product of the legitimately fertilised
flowers.* The two legitimate unions compared with

---

* II. Müller says ('Die Befruchtung,' &c., p. 352) that the
long-styled flowers, when illegitimately fertilised, yield as many
seeds as when legitimately fertilised ; but by adding up the
number of seeds from all the capsules produced by the two methods
of fertilisation, as given by him,
I arrive at the results shown in
Table 13. The average number
in the long-styled capsules, when
legitimately fertilised, is 91·4,
and when illegitimately fertilised,
77·5 ; or as 100 to 85. H. Müller
agrees with me that this is the
proper manner of viewing the
case.

the two illegitimate together yield seeds in the ratio of 100 to 61.

H. Müller also tried the effects of illegitimately fertilising the long-styled and short-styled flowers with their own pollen, instead of with that from another plant of the same form; and the results are very striking. For the capsules from the long-styled flowers thus treated contained, on an average, only 15·7 seeds instead of 77·5; and those from the short-styled 6·5, instead of 18·7 seeds per capsule. The number 6·5 agrees closely with Mr. Scott's result from the same form similarly fertilised.

From some observations by Dr. Torrey, *Hottonia inflata*, an inhabitant of the United States, does not appear to be heterostyled, but is remarkable from producing cleistogamic flowers, as will be seen in the last chapter of this volume.

Besides the genera Primula and Hottonia, *Androsace* (vel Gregoria, vel Aretia) *vitalliana* is heterostyled. Mr. Scott[*] fertilised with their own pollen 21 flowers on three short-styled plants in the Edinburgh Botanic Gardens, and not one yielded a single seed; but eight of them which were fertilised with pollen from one of the other plants of the same form, set two empty capsules. He was able to examine only dried specimens of the long-styled forms. But the evidence seems sufficient to leave hardly a doubt that Androsace is heterostyled. Fritz Müller sent me from South Brazil dried flowers of a Statice which he believed to be heterostyled. In the one form the pistil was considerably longer and the stamens slightly shorter than the corresponding organs in the other form. But as in the shorter-styled form the stigmas reached up to the anthers

---

[*] See also Treviranus in ' Bot. Zeitung,' 1863, p. 6, on this plant being dimorphic.

of the same flower, and as I could not detect in the
dried specimens of the two forms any difference in their
stigmas, or in the size of their pollen-grains, I dare not
rank this plant as heterostyled. From statements made
by Vaucher I was led to think that *Soldanella alpina* was
heterostyled, but it is impossible that Kerner, who has
closely studied this plant, could have overlooked the
fact. So again from other statements it appeared probable that Pyrola might be heterostyled, but H. Müller
examined for me two species in North Germany, and
found this not to be the case.

CHAPTER II.

Hybrid Primulas.

The Oxlip a hybrid naturally produced between Primula veris and vulgaris—The differences in structure and function between the two parent-species—Effects of crossing long-styled and short-styled Oxlips with one another and with the two forms of both parent-species—Character of the offspring from Oxlips artificially self-fertilised and cross-fertilised in a state of nature—Primula elatior shown to be a distinct species—Hybrids between other heterostyled species of Primula—Supplementary note on spontaneously produced hybrids in the genus Verbascum.

THE various species of Primula have produced in a state of nature throughout Europe an extraordinary number of hybrid forms. For instance, Professor Kerner has found no less than twenty-five such forms in the Alps.* The frequent occurrence of hybrids in this genus no doubt has been favoured by most of the species being heterostyled, and consequently requiring cross-fertilisation by insects; yet in some other genera, species which are not heterostyled and which in some respects appear not well adapted for hybrid-fertilisation, have likewise been largely hybridised. In certain districts of England, the common oxlip—a hybrid between the cowslip (*P. veris*, vel *officinalis*) and the primrose (*P. vulgaris*, vel *acaulis*)—is frequently found, and it occurs occasionally almost every-

---

* "Die Primulaceen-Bastarten," 'Oesterr. Bot. Zeitschrift,' Jahr 1875, Nos. 3, 4, and 5. See also Godron on hybrid Primulas in 'Bull. Soc. Bot. de France,' tom. x. 1853, p. 178. Also in 'Revue des Sciences Nat.' 1875, p. 331.

where. Owing to the frequency of this intermediate
hybrid form, and to the existence of the Bardfield
oxlip (*P. elatior*), which resembles to a certain extent
the common oxlip, the claim of the three forms to
rank as distinct species has been discussed oftener
and at greater length than that of almost any other
plant. Linnæus considered *P. veris, vulgaris* and
elatior to be varieties of the same species, as do some
distinguished botanists at the present day; whilst
others who have carefully studied these plants do not
doubt that they are distinct species. The following
observations prove, I think, that the latter view is
correct; and they further show that the common oxlip
is a hybrid between *P. veris* and *vulgaris*.

The cowslip differs so conspicuously in general appearance from the primrose, that nothing need here
be said with respect to their external characters.*
But some less obvious differences deserve notice. As
both species are heterostyled, their complete fertilisation depends on insects. The cowslip is habitually
visited during the day by the larger humble-bees
(viz. *Bombus muscorum* and *hortorum*), and at night
by moths, as I have seen in the case of *Cucullia*. The
primrose is never visited (and I speak after many
years' observation) by the larger humble-bees, and
only rarely by the smaller kinds; hence its fertilisation must depend almost exclusively on moths.
There is nothing in the structure of the flowers of the
two plants which can determine the visits of such
widely different insects. But they emit a different
odour, and perhaps their nectar may have a different
taste. Both the long-styled and short-styled forms of

---

* The Rev. W. A. Leighton
has pointed out certain differences
in the form of the capsules and
seed, in 'Ann. and Mag. of Nat.
Hist.' 2nd series, vol. ii. 1848,
p. 164.

the primrose, when legitimately and naturally fertilised, yield on an average many more seeds per capsule
than the cowslip, namely, in the proportion of 100 to 55.
When illegitimately fertilised they are likewise more
fertile than the two forms of the cowslip, as shown by
the larger proportion of their flowers which set capsules, and by the larger average number of seeds which
the capsules contain. The difference also between the
number of seeds produced by the long-styled and short-
styled flowers of the primrose, when both are illegitimately fertilised, is greater than that between the
number produced under similar circumstances by the
two forms of the cowslip. The long-styled flowers of
the primrose when protected from the access of all insects, except such minute ones as Thrips, yield a considerable number of capsules containing on an average
19·2 seeds per capsule ; whereas 18 plants of the long-
styled cowslip similarly treated did not yield a single
seed.

The primrose, as every one knows, flowers a little
earlier in the spring than the cowslip, and inhabits
slightly different stations and districts. The primrose
generally grows on banks or in woods, whilst the cowslip is found in more open places. The geographical
range of the two forms is different. Dr. Bromfield remarks * that " the primrose is absent from all the interior region of northern Europe, where the cowslip is
indigenous." In Norway, however, both plants range
to the same degree of north latitude.†

The cowslip and primrose, when intercrossed, be-

---

* 'Phytologist,' vol. iii. p. 694.
† H. Lecoq, 'Géograph. Bot. de
l'Europe,' tom. viii. 1858, pp. 141,
144. See also 'Ann. and Mag. of
Nat. Hist.' ix. 1842, pp. 156,
515. Also Boreau, 'Flore du
centre de la France,' 1840, tom. ii.
p. 376. With respect to the rarity
of *P. veris* in western Scotland,
see H. C. Watson, 'Cybele Britannica,' ii. p. 293.

have like distinct species, for they are far from
being mutually fertile. Gärtner* crossed 27 flowers
of *P. vulgaris* with pollen of *P. veris*, and obtained
16 capsules; but these did not contain any good
seed. He also crossed 21 flowers of *P. veris* with
pollen of *P. vulgaris*; and now he got only five
capsules, containing seed in a still less perfect
condition. Gärtner knew nothing about heterostylism; and his complete failure may perhaps be
accounted for by his having crossed together the
same forms of the cowslip and primrose; for such
crosses would have been of an illegitimate as well as
of a hybrid nature, and this would have increased
their sterility. My trials were rather more fortunate.
Twenty-one flowers, consisting of both forms of the
cowslip and primrose, were intercrossed legitimately,
and yielded seven capsules (*i.e.* 33 per cent.), containing on an average 42 seeds; some of these seeds,
however, were so poor that they probably would not have
germinated. Twenty-one flowers on the same cowslip
and primrose plants were also intercrossed illegitimately, and they likewise yielded seven capsules (or
33 per cent.), but these contained on an average only
13 good and bad seeds. I should, however, state that
some of the above flowers of the primrose were fertilised
with pollen from the polyanthus, which is certainly a
variety of the cowslip, as may be inferred from the perfect fertility *inter se* of the crossed offspring from these
two plants.† To show how sterile these hybrid unions

---

* 'Bastarderzeugung,' 1849, p.
721.

† Mr. Scott has discussed the
nature of the polyanthus ('Proc.
Linn. Soc.' viii. Bot. 1864, p.
103), and arrives at a different
conclusion; but I do not think
that his experiments were suffi-

ciently numerous. The degree of
infertility of a cross is liable to
much fluctuation. Pollen from
the cowslip at first appears rather
more efficient on the primrose than
that of the polyanthus; for 12
flowers of both forms of the primrose, fertilised legitimately and

were I may remind the reader that 90 per cent. of the
flowers of the primrose fertilised legitimately with
primrose-pollen yielded capsules, containing on an
average 66 seeds; and that 54 per cent. of the flowers
fertilised illegitimately yielded capsules containing on
an average 35·5 seeds per capsule. The primrose,
especially the short-styled form, when fertilised by the
cowslip, is less sterile, as Gärtner likewise observed,
than is the cowslip when fertilised by the primrose. The
above experiments also show that a cross between the
same forms of the primrose and cowslip is much more
sterile than that between different forms of these two
species.

The seeds from the several foregoing crosses were
sown, but none germinated except those from the
short-styled primrose fertilised with pollen of the
polyanthus; and these seeds were the finest of the
whole lot. I thus raised six plants, and compared
them with a group of wild oxlips which I had transplanted into my garden. One of these wild oxlips
produced slightly larger flowers than the others, and
this one was identical in every character (in foliage,
flower-peduncle, and flowers) with my six plants,
excepting that the flowers of the latter were tinged of
a dingy red colour, from being descended from the
polyanthus.

We thus see that the cowslip and primrose cannot be crossed either way except with considerable
difficulty, that they differ conspicuously in external
appearance, that they differ in various physiological

---

illegitimately with pollen of the
cowslip gave five capsules, containing on an average 32·4 seeds;
whilst 18 flowers similarly fertilised by polyanthus-pollen yielded
only five capsules, containing only
22·6 seeds. On the other hand,
the seeds produced by the polyanthus-pollen were much the
finest of the whole lot, and were
the only ones which germinated.

characters, that they inhabit slightly different stations and range differently. Hence those botanists who rank these plants as varieties ought to be able to prove that they are not as well fixed in character as are most species; and the evidence in favour of such instability of character appears at first sight very strong. It rests, first, on statements made by several competent observers that they have raised cowslips, primroses, and oxlips from seeds of the same plant; and, secondly, on the frequent occurrence in a state of nature of plants presenting every intermediate gradation between the cowslip and primrose.

The first statement, however, is of little value; for, heterostylism not being formerly understood, the seed-bearing plants were in no instance* protected from the visits of insects; and there would be almost as much risk of an isolated cowslip, or of several cowslips if consisting of the same form, being crossed by a neighbouring primrose and producing oxlips, as of one sex of a diœcious plant, under similar circumstances, being crossed by the opposite sex of an allied and neighbouring species. Mr. H. C. Watson, a critical and most careful observer, made many experiments by sowing the seeds of cowslips and of various kinds of oxlips, and arrived at the following conclusion,† namely, "that seeds of a cowslip can produce cowslips and oxlips, and that seeds of an oxlip can produce cowslips, oxlips, and primroses." This conclusion harmonises perfectly with the view that in

* One author states in the 'Phytologist' (vol. iii. p. 703) that he covered with bell-glasses some cowslips, primroses, &c., on which he experimented. He specifies all the details of his experiment, but does not say that he artificially fertilised his plants; yet he obtained an abundance of seed, which is simply impossible. Hence there must have been some strange error in these experiments, which may be passed over as valueless.

† 'Phytologist,' ii. pp. 217, 852; iii. p. 43.

all cases, when such results have been obtained, the
unprotected cowslips have been crossed by primroses,
and the unprotected oxlips by either cowslips or
primroses; for in this latter case we might expect, by
the aid of reversion, which notoriously comes into
powerful action with hybrids, that the two parent-forms
in appearance pure, as well as many intermediate gradations, would be occasionally produced. Nevertheless
the two following statements offer considerable difficulty. The Rev. Prof. Henslow* raised from seeds of a
cowslip growing in his garden, various kinds of oxlips
and one perfect primrose; but a statement in the same
paper perhaps throws light on this anomalous result.
Prof. Henslow had previously transplanted into his
garden a cowslip, which completely changed its appearance during the following year, and now resembled
an oxlip. Next year again it changed its character,
and produced, in addition to the ordinary umbels, a
few single-flowered scapes, bearing flowers somewhat
smaller and more deeply coloured than those of the
common primrose. From what I have myself observed
with oxlips, I cannot doubt that this plant was an oxlip in a highly variable condition, almost like that of
the famous *Cytisus adami.* This presumed oxlip was
propagated by offsets, which were planted in different
parts of the garden; and if Prof. Henslow took by
mistake seeds from one of these plants, especially if it
had been crossed by a primrose, the result would be
quite intelligible. Another case is still more difficult
to understand: Dr. Herbert† raised, from the seeds of
a highly cultivated red cowslip, cowslips, oxlips of
various kinds, and a primrose. This case, if accurately

---

* Loudon's 'Mag. of Nat. Hist.' iii. 1830, p. 409.
† 'Transact. Hort. Soc.' iv. p. 19.

recorded, which I much doubt, is explicable only on
the improbable assumption that the red cowslip was
not of pure parentage. With species and varieties
of many kinds, when intercrossed, one is sometimes
strongly prepotent over the other; and instances are
known* of a variety crossed by another, producing
offspring which in certain characters, as in colour,
hairiness, &c., have proved identical with the pollen-
bearing parent, and quite dissimilar to the mother-
plant; but I do not know of any instance of the offspring of a cross perfectly resembling, in a considerable number of important characters, the father alone.
It is, therefore, very improbable that a pure cowslip
crossed by a primrose should ever produce a primrose
in appearance pure.   Although the facts given by Dr.
Herbert and Prof. Henslow are difficult to explain, yet
until it can be shown that a cowslip or a primrose,
carefully protected from insects, will give birth to at
least oxlips, the cases hitherto recorded have little
weight in leading us to admit that the cowslip and
primrose are varieties of one and the same species.

Negative evidence is of little value; but the following facts may be worth giving :—Some cowslips which
had been transplanted from the fields into a shrubbery
were again transplanted into highly manured land.
In the following year they were protected from insects,
artificially fertilised, and the seed thus procured was
sown in a hotbed. The young plants were afterwards
planted out, some in very rich soil, some in stiff poor
clay, some in old peat, and some in pots in the green-
house; so that these plants, 765 in number, as well as
their parents, were subjected to diversified and un-

---

* I have given instances in my
work 'On the Variation of Animals and Plants under Domestication,' chap. xv. 2nd edit. vol.
ii. p. 69.

natural treatment; but not one of them presented the least variation except in size—those in the peat attaining almost gigantic dimensions, and those in the clay being much dwarfed.

I do not, of course, doubt that cowslips exposed during *several* successive generations to changed conditions would vary, and that this might occasionally occur in a state of nature. Moreover, from the law of analogical variation, the varieties of any one species of Primula would probably in some cases resemble other species of the genus. For instance I raised a red primrose from seed from a protected plant, and the flowers, though still resembling those of the primrose, were borne during one season in umbels on a long footstalk like that of a cowslip.

With regard to the second class of facts in support of the cowslip and primrose being ranked as mere varieties, namely, the well-ascertained existence in a state of nature of numerous linking forms* :—If it can be shown that the common wild oxlip, which is intermediate in character between the cowslip and primrose, resembles in sterility and other essential respects a hybrid plant, and if it can further be shown that the oxlip, though in a high degree sterile, can be fertilised by either parent-species, thus giving rise to still finer gradational links, then the presence of such linking forms in a state of nature ceases to be an argument of any weight in favour of the cowslip and primrose being varieties, and becomes, in fact, an argument on the other side. The hybrid origin of a plant in a state of nature can be recognised by four tests: first, by its occurrence only where both presumed parent-

* See an excellent article on this subject by Mr. H. C. Watson in the 'Phytologist,' vol. iii. p. 43.

species exist or have recently existed; and this holds good, as far as I can discover, with the oxlip; but the *P. elatior* of Jacq., which, as we shall presently see, constitutes a distinct species, must not be confounded with the common oxlip. Secondly, by the supposed hybrid plant being nearly intermediate in character between the two parent-species, and especially by its resembling hybrids artificially made between the same two species. Now the oxlip is intermediate in character, and resembles in every respect, except in the colour of the corolla, hybrids artificially produced between the primrose and the polyanthus, which latter is a variety of the cowslip. Thirdly, by the supposed hybrids being more or less sterile when crossed *inter se*: but to try this fairly two distinct plants of the same parentage, and not two flowers on the same plant, should be crossed; for many pure species are more or less sterile with pollen from the same individual plant; and in the case of hybrids from heterostyled species the opposite forms should be crossed. Fourthly and lastly, by the supposed hybrids being much more fertile when crossed with either pure parent-species than when crossed *inter se*, but still not as fully fertile as the parent-species.

For the sake of ascertaining the two latter points, I transplanted a group of wild oxlips into my garden. They consisted of one long-styled and three short-styled plants, which, except in the corolla of one being slightly larger, resembled each other closely. The trials which were made, and the results obtained, are shown in the five following tables. No less than twenty different crosses are necessary in order to ascertain fully the fertility of hybrid heterostyled plants, both *inter se* and with their two parent-species. In this instance 256 flowers

were crossed in the course of four seasons. I may
mention, as a mere curiosity, that if any one were to
raise hybrids between two trimorphic heterostyled
species, he would have to make 90 distinct unions
in order to ascertain their fertility in all ways;
and as he would have to try at least 10 flowers in
each case, he would be compelled to fertilise 900
flowers and count their seeds. This would probably
exhaust the patience of the most patient man.

TABLE 14.

Crosses inter se *between the two forms of the common Oxlip.*

Illegitimate union.	Legitimate union.	Illegitimate union.	Legitimate union.
Short-styled ox-lip, by pollen of short-styled oxlip: 20 flowers fertilised, did not produce one capsule.	Short-styled ox-lip, by pollen of long-styled oxlip: 10 flowers fertilised, did not produce one capsule.	Long-styled ox-lip, by its own pollen: 24 flowers fertilised, produced five capsules, containing 6, 10, 20, 8, and 14 seeds. Average 11·6.	Long-styled ox-lip, by pollen o short-styled oxlip 10 flowers fertilised did not produce on capsule.

TABLE 15.

Both forms of the Oxlip crossed with Pollen of both forms of the Cowslip, P. veris.

Illegitimate union.	Legitimate union.	Illegitimate union.	Legitimate union.
Short-styled ox-lip, by pollen of short-styled cowslip: 18 flowers fertilised, did not produce one capsule.	Short-styled ox-lip, by pollen of long-styled cowslip: 18 flowers fertilised, produced three capsules, containing 7, 3, and 3 wretched seeds, apparently incapable of germination.	Long-styled ox-lip, by pollen of long-styled cowslip: 11 flowers fertilised, produced one capsule, containing 13 wretched seeds.	Long-styled ox-lip, by pollen of short-styled cowslip: 5 flowers fertilised, produced two capsules, containing 21 and 28 very fine seeds.

F

Table 16.

Both forms of the Oxlip crossed with Pollen of both forms of the Primrose, P. vulgaris.

Illegitimate union.	Legitimate union.	Illegitimate union.	Legitimate union.
Short-styled ox-lip, by pollen of short-styled prim-rose: 34 flowers fertilised, produced two capsules, containing 5 and 12 seeds.	Short-styled ox-lip, by pollen of long-styled prim-rose: 26 flowers fertilised, produced six capsules, containing 16, 20, 5, 10, 19, and 24 seeds. Average 15·7. Many of the seeds very poor, some good.	Long-styled ox-lip, by pollen of long-styled prim-rose: 11 flowers fertilised, produced four capsules, containing 10, 7. 5, and 6 wretched seeds. Average 7·0.	Long-styled ox-lip, by pollen of short-styled prim-rose: 5 flowers fertilised, produced five capsules, containing 26, 32, 23, 28, and 34 seeds. Average 28·6.

Table 17.

Both forms of the Cowslip crossed with Pollen of both forms of the Oxlip.

Illegitimate union.	Legitimate union.	Illegitimate union.	Legitimate union.
Short-styled cow-slip, by pollen of short-styled oxlip: 8 flowers fertilised, produced not one capsule.	Long-styled cow-slip, by pollen of short-styled oxlip: 8 flowers fertilised, produced one capsule, containing 26 seeds.	Long-styled cow-slip, by pollen of long-styled oxlip: 8 flowers fertilised, produced three capsules, containing 5, 6, and 14 seeds. Average 8·3.	Short-styled cow-slip, by pollen of long-styled oxlip: 8 flowers fertilised, produced eight capsules, containing 58, 38, 31, 44, 23, 26, 37, and 66 seeds. Average 40·4.

Table 18.

Both forms of the Primrose crossed with Pollen of both forms of the Oxlip.

Illegitimate union.	Legitimate union.	Illegitimate union.	Legitimate union.
Short-styled prim-rose, by pollen of short-styled oxlip: 8 flowers fertilised, produced not one capsule.	Long-styled prim-rose, by pollen of short-styled oxlip: 8 flowers fertilised, produced two capsules, containing 5 and 2 seeds.	Long-styled prim-rose, by pollen of long-styled oxlip: 8 flowers fertilised, produced eight capsules, containing 15, 7, 12, 20, 22, 7, 16, and 13 seeds. Average 14·0.	Short-styled prim-rose, by pollen of long-styled oxlip: 8 flowers fertilised, produced four capsules, containing 52, 52, 42, and 49 seeds, some good and some bad. Average 48·7.

We see in these five tables the number of capsules and of seeds produced, by crossing both forms of the oxlip in a legitimate and illegitimate manner with one another, and with the two forms of the primrose and cowslip. I may premise that the pollen of two of the short-styled oxlips consisted of nothing but minute aborted whitish cells; but in the third short-styled plant about one-fifth of the grains appeared in a sound condition. Hence it is not surprising that neither the short-styled nor the long-styled oxlip produced a single seed when fertilised with this pollen. Nor did the pure cowslips or primroses when illegitimately fertilised with it; but when thus legitimately fertilised they yielded a few good seeds. The female organs of the short-styled oxlips, though greatly deteriorated in power, were in a rather better condition than the male organs; for though the short-styled oxlips yielded no seed when fertilised by the long-styled oxlips, and hardly any when illegitimately fertilised by pure cowslips or primroses, yet when legitimately fertilised by these latter species, especially by the long-styled primrose, they yielded a moderate supply of good seed.

The long-styled oxlip was more fertile than the three short-styled oxlips, and about half its pollen-grains appeared sound. It bore no seed when legitimately fertilised by the short-styled oxlips; but this no doubt was due to the badness of the pollen of the latter; for when illegitimately fertilised (Table 14) by its own pollen it produced some good seeds, though much fewer than self-fertilised cowslips or primroses would have produced. The long-styled oxlip likewise yielded a very low average of seed, as may be seen in the third compartment of the four latter tables, when illegitimately fertilised by, and when

illegitimately fertilising, pure cowslips and primroses.
The four corresponding legitimate unions, however, were
moderately fertile, and one (viz. that between a short-
styled cowslip and the long-styled oxlip in Table 17)
was nearly as fertile as if both parents had been pure.
A short-styled primrose legitimately fertilised by the
long-styled oxlip (Table 18) also yielded a moderately
good average, namely 48·7 seeds; but if this short-
styled primrose had been fertilised by a long-styled
primrose it would have yielded an average of 65 seeds.
If we take the ten legitimate unions together, and the
ten illegitimate unions together, we shall find that 29
per cent. of the flowers fertilised in a legitimate manner
yielded capsules, these containing on an average 27·4
good and bad seeds; whilst only 15 per cent. of the
flowers fertilised in an illegitimate manner yielded
capsules, these containing on an average only 11·0
good and bad seeds.

In a previous part of this chapter it was shown that
illegitimate crosses between the long-styled form of
the primrose and the long-styled cowslip, and between
the short-styled primrose and short-styled cowslip, are
more sterile than legitimate crosses between these two
species; and we now see that the same rule holds good
almost invariably with their hybrid offspring, whether
these are crossed *inter se*, or with either parent-species;
so that in this particular case, but not as we shall presently see in other cases, the same rule prevails with
the pure unions between the two forms of the same
heterostyled species, with crosses between two distinct
heterostyled species, and with their hybrid offspring.

Seeds from the long-styled oxlip fertilised by its
own pollen were sown, and three long-styled plants
raised. The first of these was identical in every
character with its parent. The second bore rather

smaller flowers, of a paler colour, almost like those of
the primrose; the scapes were at first single-flowered,
but later in the season a tall thick scape, bearing many
flowers, like that of the parent oxlip, was thrown up.
The third plant likewise produced at first only single-
flowered scapes, with the flowers rather small and of a
darker yellow; but it perished early. The second
plant also died in September; and the first plant,
though all three grew under very favourable conditions, looked very sickly. Hence we may infer that
seedlings from self-fertilised oxlips would hardly be
able to exist in a state of nature. I was surprised to
find that all the pollen-grains in the first of these seedling oxlips appeared sound; and in the second only a
moderate number were bad. These two plants, however,
had not the power of producing a proper number of
seeds; for though left uncovered and surrounded by
pure primroses and cowslips, the capsules were estimated to include an average of only from fifteen to
twenty seeds.

From having many experiments in hand, I did not
sow the seed obtained by crossing both forms of the
primrose and cowslip with both forms of the oxlip,
which I now regret; but I ascertained an interesting point, namely, the character of the offspring
from oxlips growing in a state of nature near both
primroses and cowslips. The oxlips were the same
plants which, after their seeds had been collected, were
transplanted and experimented on. From the seeds
thus obtained eight plants were raised, which, when
they flowered, might have been mistaken for pure
primroses; but on close comparison the eye in the
centre of the corolla was seen to be of a darker yellow,
and the peduncles more elongated. As the season ad-
vanced, one of these plants threw up two naked scapes,

7 inches in height, which bore umbels of flowers of
the same character as before.   This fact led me to examine the other plants after they had flowered and
were dug up; and I found that the flower-peduncles
of all sprung from an extremely short common scape,
of which no trace can be found in the pure primrose.
Hence these plants are beautifully intermediate between the oxlip and the primrose, inclining rather
towards the latter; and we may safely conclude that the
parent oxlips had been fertilised by the surrounding
primroses.

From the various facts now given, there can be no
doubt that the common oxlip is a hybrid between the
cowslip (*P. veris*, Brit. Fl.) and the primrose (*P. vulgaris*, Brit. Fl.), as has been surmised by several
botanists.   It is probable that oxlips may be produced
either from the cowslip or the primrose as the seed-
bearer, but oftenest from the latter, as I judge from
the nature of the stations in which oxlips are generally
found,* and from the primrose when crossed by the
cowslip being more fertile than, conversely, the cowslip
by the primrose.   The hybrids themselves are also
rather more fertile when crossed with the primrose
than with the cowslip.   Whichever may be the seed-
bearing plant, the cross is probably between different
forms of the two species; for we have seen that legitimate hybrid unions are more fertile than illegitimate
hybrid unions.   Moreover a friend in Surrey found
that 29 oxlips which grew in the neighbourhood of
his house consisted of 13 long-styled and 16 short-
styled plants; now, if the parent-plants had been
illegitimately united, either the long- or short-styled
form would have greatly preponderated, as we shall

* See also on this head Hardwicke's 'Science Gossip,' 1867, pp.
114, 137.

hereafter see good reason to believe. The case of
the oxlip is interesting; for hardly any other instance is known of a hybrid spontaneously arising
in such large numbers over so wide an extent of country. The common oxlip (not the *P. elatior* of Jacq.)
is found almost everywhere throughout England, where
both cowslips and primroses grow. In some districts,
as I have seen near Hartfield in Sussex and in parts
of Surrey, specimens may be found on the borders of
almost every field and small wood. In other districts
the oxlip is comparatively rare: near my own resi-
dence I have found, during the last twenty-five years,
not more than five or six plants or groups of plants.
It is difficult to conjecture what is the cause of this
difference in their number. It is almost necessary
that a plant, or several plants belonging to the same
form, of one parent-species, should grow near the
opposite form of the other parent-species; and it is
further necessary that both species should be frequented by the same kind of insect, no doubt a moth.
The cause of the rare appearance of the oxlip in
certain districts may be the rarity of some moth,
which in other districts habitually visits both the
primrose and cowslip.

Finally, as the cowslip and primrose differ in the
various characters above specified,—as they are in a
high degree sterile when intercrossed,—as there is no
trustworthy evidence that either species, when un-
crossed, has ever given birth to the other species or
to any intermediate form,—and as the intermediate
forms which are often found in a state of nature have
been shown to be more or less sterile hybrids of the
first or second generation,—we must for the future
look at the cowslip and primrose as good and true
species.

Primula elatior, Jacq., or the Bardfield Oxlip, is
found in England only in two or three of the eastern
counties. On the Continent it has a somewhat different range from that of the cowslip and primrose;
and it inhabits some districts where neither of these
species live.* In general appearance it differs so
much from the common oxlip, that no one accustomed
to see both forms in the living state could afterwards
confound them; but there is scarcely more than a
single character by which they can be distinctly defined, namely, their linear-oblong capsules equalling
the calyx in length.† The capsules when mature differ
possible methods, they behave like the other heterostyled species of the genus, but differ somewhat (see
Tables 8 and 12) in the smaller proportion of the illegitimately fertilised flowers which set capsules. That
P. elatior is not a hybrid is certain, for when the two
forms were legitimately united they yielded the large
average of 47·1 seeds, and when illegitimately united
35·5 per capsule; whereas, of the four possible unions
(Table 14) between the two forms of the common oxlip which we know to be a hybrid, one alone yielded
any seed; and in this case the average number was
only 11·6 per capsule. Moreover I could not detect
a single bad pollen-grain in the anthers of the short-
styled *P. elatior;* whilst in two short-styled plants of
the common oxlip all the grains were bad, as were
a large majority in a third plant. As the common

---

* For England, see Hewett C.
Watson, 'Cybele Britannica,' vol.
ii. 1849, p. 292. For the Continent, see Lecoq, ' Géograph.
Botanique de l'Europe,' tom. viii.

1858, p. 142. For the Alps, see
' Ann. and Mag. Nat. Hist.' vol.
ix. 1842, pp. 156 and 515.
   † Babington's ' Manual of British Botany,' 1851, p. 258.

oxlip is a hybrid between the primrose and cowslip, it is not surprising that eight long-styled flowers of the primrose, fertilised by pollen from the long-styled common oxlip, produced eight capsules (Table 18), containing, however, only a low average of seeds; whilst the same number of flowers of the primrose, similarly fertilised by the long-styled Bardfield oxlip, produced only a single capsule; this latter plant being an altogether distinct species from the primrose. Plants of *P. elatior* have been propagated by seed in a garden for twenty-five years, and have kept all this time quite constant, excepting that in some cases the flowers varied a little in size and tint.[*] Nevertheless, according to Mr. H. C. Watson and Dr. Bromfield,[†] plants may be occasionally found in a state of nature, in which most of the characters by which this species can be distinguished from *P. veris* and *vulgaris* fail; but such intermediate forms are probably due to hybridisation; for Kerner states, in the paper before referred to, that hybrids sometimes, though rarely, arise in the Alps between *P. elatior* and *veris*.

Finally, although we may freely admit that *Primula veris*, *vulgaris*, and *elatior*, as well as all the other species of the genus, are descended from a common primordial form, yet from the facts above given, we must conclude that these three forms are now as fixed in character as are many others which are universally ranked as true species. Consequently they have as good a right to receive distinct specific names as have, for instance, the ass, quagga, and zebra.

Mr. Scott has arrived at some interesting results by

---

[*] See Mr. H. Doubleday in the 'Gardener's Chronicle,' 1867, p. 435, also Mr. W. Marshall, ibid.   p. 462.
[†] 'Phytologist,' vol. i. p. 1001, and vol. iii. p. 695.

crossing other heterostyled species of Primula.* I
have already alluded to his statement, that in four
instances (not to mention others) a species when crossed
with a distinct one yielded a larger number of seeds
than the same species fertilised illegitimately with its
own-form pollen, though taken from a distinct plant.
It has long been known from the researches of Kölreuter
and Gärtner, that two species when crossed reciprocally
sometimes differ as widely as is possible in their fertility : thus A when crossed with the pollen of B will
yield a large number of seeds, whilst B may be crossed
repeatedly with pollen of A, and will never yield a single
seed.   Now Mr. Scott shows in several cases that the
same law holds good when two heterostyled species
of Primula are intercrossed, or when one is crossed
with a homostyled species.   But the results are much
more complicated than with ordinary plants, as two
heterostyled dimorphic species can be intercrossed in
eight different ways.   I will give one instance from
Mr. Scott.   The long-styled *P. hirsuta* fertilised legitimately and illegitimately with pollen from the two
forms of *P. auricula*, and reciprocally the long-styled
P. auricula fertilised legitimately and illegitimately
with pollen from the two forms of *P. hirsuta*, did
not produce a single seed.   Nor did the short-
styled *P. hirsuta* when fertilised legitimately and
illegitimately with the pollen of the two forms of
P. auricula.   On the other hand, the short-styled
P. auricula fertilised with pollen from the long-styled
P. hirsuta yielded capsules containing on an average
no less than 56 seeds ; and the short-styled *P.
auricula* by pollen of the short-styled *P. hirsuta*
yielded capsules containing on an average 42 seeds

---

* 'Journ. Linn. Soc. Bot.' vol. viii. 1864, p. 93 to end.

per capsule. So that out of the eight possible unions
between the two forms of these two species, six
were utterly barren, and two fairly fertile. We have
seen also the same sort of extraordinary irregularity in
the results of my twenty different crosses (Tables
14 to 18), between the two forms of the oxlip, primrose, and cowslip. Mr. Scott remarks, with respect
to the results of his trials, that they are very surprising,
as they show us that " the sexual forms of a species
manifest in their respective powers for conjunction
with those of another species, physiological peculiarities which might well entitle them, by the criterion of
fertility, to specific distinction."

Finally, although *P. veris* and *vulgaris*, when crossed
legitimately, and especially when their hybrid offspring
are crossed in this manner with both parent-species,
were decidedly more fertile, than when crossed in an
illegitimate manner, and although the legitimate cross
effected by Mr. Scott between *P. auricula* and *hirsuta*
was more fertile, in the ratio of 56 to 42, than the
illegitimate cross, nevertheless it is very doubtful,
from the extreme irregularity of the results in the
various other hybrid crosses made by Mr. Scott, whether
it can be predicted that two heterostyled species are
generally more fertile if crossed legitimately (*i.e.* when
opposite forms are united) than when crossed illegitimately.

Supplementary Note on some wild hybrid Verbascums.

In an early part of this chapter I remarked that few
other instances could be given of a hybrid spontaneously arising in such large numbers, and over so wide an
extent of country, as that of the common oxlip; but perhaps the number of well-ascertained cases of naturally

produced hybrid willows is equally great.* Numerous spontaneous hybrids between several species of Cistus, found near Narbonne, have been carefully described by M. Timbal-Lagrave,† and many hybrids between an Aceras and Orchis have been observed by Dr. Weddell.‡ In the genus Verbascum, hybrids are supposed to have often originated§ in a state of nature; some of these undoubtedly are hybrids, and several hybrids have originated in gardens; but most of these cases require,‖ as Gärtner remarks, verification. Hence the following case is worth recording, more especially as the two species in question, *V. thapsus* and *lychnitis,* are perfectly fertile when insects are excluded, showing that the stigma of each flower receives its own pollen. Moreover the flowers offer only pollen to insects, and have not been rendered attractive to them by secreting nectar.

I transplanted a young wild plant into my garden for experimental purposes, and when it flowered it plainly differed from the two species just mentioned and from a third which grows in this neighbourhood. I thought that it was a strange variety of *V. thapsus.* It attained the height (by measurement) of 8 feet! It was covered with a net, and ten flowers were fertilised with pollen from the same plant; later in the season, when uncovered, the flowers were freely visited by pollen-collecting bees; nevertheless, although many capsules were produced, not one contained a single seed. During the following year this same plant was

---

* Max Wichura, 'Die Bastard-befruchtung, &c., der Weiden,' 1865.

† 'Mém. de l'Acad. des Sciences de Toulouse,' 5ᵉ série, tom. v. p. 28.

‡ 'Annales des Sc. Nat.' 3ᵉ série, Bot. tom. xviii. p. 6.

§ See, for instance, the 'English Flora,' by Sir J. E. Smith, 1824, vol. i. p. 307.

‖ See Gärtner, 'Bastarderzeugung,' 1849, p. 590.

left uncovered near plants of *V. thapsus* and *lychnitis ;*
but again it did not produce a single seed. Four
flowers, however, which were repeatedly fertilised
with pollen of *V. lychnitis*, whilst the plant was tem-
porarily kept under a net, produced four capsules,
which contained five, one, two, and two seeds ; at the
same time three flowers were fertilised with pollen of
V. thapsus, and these produced two, two, and three
seeds. To show how unproductive these seven capsules
were, I may state that a fine capsule from a plant of
V. thapsus growing close by contained above 700 seeds.
These facts led me to search the moderately-sized field
whence my plant had been removed, and I found in it
many plants of *V. thapsus* and *lychnitis* as well as
thirty-three plants intermediate in character between
these two species. These thirty-three plants differed
much from one another. In the branching of the stem
they more closely resembled *V. lychnitis* than *V. thapsus*,
but in height the latter species. In the shape of their
leaves they often closely approached *V. lychnitis*, but
some had leaves extremely woolly on the upper surface
and decurrent like those of *V. thapsus ;* yet the degree
of woolliness and of decurrency did not always go
together. In the petals being flat and remaining
open, and in the manner in which the anthers of the
longer stamens were attached to the filaments, these
plants all took more after *V. lychnitis* than *V. thapsus*.
In the yellow colour of the corolla they all resembled
the latter species. On the whole, these plants appeared
to take rather more after *V. lychnitis* than *V. thapsus*.
On the supposition that they were hybrids, it is not an
anomalous circumstance that they should all have produced yellow flowers ; for Gärtner crossed white and
yellow-flowered varieties of Verbascum, and the offspring thus produced never bore flowers of an inter-

mediate tint, but either pure white or pure yellow
flowers, generally of the latter colour.*

My observations were made in the autumn; so that
I was able to collect some half-matured capsules from
twenty of the thirty-three intermediate plants, and
likewise capsules of the pure *V. lychnitis* and *thapsus*
growing in the same field. All the latter were filled
with perfect but immature seeds, whilst the capsules of
the twenty intermediate plants did not contain one
single perfect seed. These plants, consequently, were
absolutely barren. From this fact,—from the one plant
which was transplanted into my garden yielding when
artificially fertilised with pollen from *V. lychnitis* and
thapsus some seeds, though extremely few in number,—
from the circumstance of the two pure species growing
in the same field,—and from the intermediate character
of the sterile plants, there can be no doubt that they
were hybrids. Judging from the position in which
they were chiefly found, I am inclined to believe they
were descended from *V. thapsus* as the seed-bearer, and
V. lychnitis as the pollen-bearer.

It is known that many species of Verbascum, when
the stem is jarred or struck by a stick, cast off their
flowers.† This occurs with *V. thapsus*, as I have repeatedly observed. The corolla first separates from its
attachment, and then the sepals spontaneously bend
inwards so as to clasp the ovarium, pushing off the
corolla by their movement, in the course of two or
three minutes. Nothing of this kind takes place with
young barely expanded flowers. With *Verbascum lychnitis* and, as I believe, *V. phœniceum* the corolla is not cast

---

* 'Bastarderzeugung,' p. 307.
† This was first observed by
Correa de Serra: see Sir J. E.
Smith's 'English Flora,' 1824, vol.
i. p. 311; also 'Life of Sir J. E.
Smith,' vol. ii. p. 210. I was
guided to these references by the
Rev. W. A. Leighton, who ob-
served this same phenomenon with
V. virgatum.

off, however often and severely the stem may be struck. In this curious property the above-described hybrids took after *V thapsus;* for I observed, to my surprise, that when I pulled off the flower-buds round the flowers which I wished to mark with a thread, the slight jar invariably caused the corollas to fall off.

These hybrids are interesting under several points of view. First, from the number found in various parts of the same moderately-sized field. That they owed their origin to insects flying from flower to flower, whilst collecting pollen, there can be no doubt. Although insects thus rob the flowers of a most precious substance, yet they do great good; for, as I have elsewhere shown,* the seedlings of *V. thapsus* raised from flowers fertilised with pollen from another plant, are more vigorous than those raised from self-fertilised flowers. But in this particular instance the insects did great harm, as they led to the production of utterly barren plants. Secondly, these hybrids are remarkable from differing much from one another in many of their characters; for hybrids of the first generation, if raised from uncultivated plants, are generally uniform in character. That these hybrids belonged to the first generation we may safely conclude, from the absolute sterility of all those observed by me in a state of nature and of the one plant in my garden, excepting when artificially and repeatedly fertilised with pure pollen, and then the number of seeds produced was extremely small. As these hybrids varied so much, an almost perfectly graduated series of forms, connecting together the two widely distinct parent-species, could easily have been selected. This case, like that of the common oxlip, shows that botanists ought to be

---

* ' The Effects of Cross and Self-fertilisation,' 1876, p. 89.

cautious in inferring the specific identity of two forms from the presence of intermediate gradations; nor would it be easy in the many cases in which hybrids are moderately fertile to detect a slight degree of sterility in such plants growing in a state of nature and liable to be fertilised by either parent-species. Thirdly and lastly, these hybrids offer an excellent illustration of a statement made by that admirable observer Gärtner, namely, that although plants which can be crossed with ease generally produce fairly fertile offspring, yet well-pronounced exceptions to this rule occur; and here we have two species of Verbascum which evidently cross with the greatest ease, but produce hybrids which are excessively sterile.

CHAPTER III.

Heterostyled Dimorphic Plants—*continued*.

Linum grandiflorum, long-styled form utterly sterile with own-form pollen—Linum perenne, torsion of the pistils in the long-styled form alone—Homostyled species of Linum—Pulmonaria officinalis, singular difference in self-fertility between the English and German long-styled plants—Pulmonaria angustifolia shown to be a distinct species, long-styled form completely self-sterile—Polygonum fagopyrum—Various other heterostyled genera—Rubiaceæ—Mitchella repens, fertility of the flowers in pairs—Houstonia—Faramea, remarkable difference in the pollen-grains of the two forms; torsion of the stamens in the short-styled form alone; development not as yet perfect—The heterostyled structure in the several Rubiaceous genera not due to descent in common.

It has long been known* that several species of Linum present two forms, and having observed this fact in *L. flavum* more than thirty years ago, I was led, after ascertaining the nature of heterostylism in Primula, to examine the first species of Linum which I met with, namely, the beautiful *L. grandiflorum.* This plant exists under two forms, occurring in about equal numbers, which differ little in structure, but greatly in function. The foliage, corolla, stamens, and pollen-grains (the latter examined both distended with water and dry) are alike in the two forms (Fig. 4). The difference is confined to the pistil; in the short-styled form the styles and the stigmas are only about half the length of those in the long-styled. A more

---

* Treviranus has shown that this is the case in his review of my original paper, 'Bot. Zeitung,' 1863, p. 189.

important distinction is, that the five stigmas in the short-styled form diverge greatly from one another, and pass out between the filaments of the stamens,

Fig. 4.

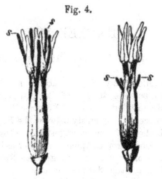

Long-styled form.   Short-styled form.
s s stigmas.
LINUM GRANDIFLORUM.

and thus lie within the tube of the corolla. In the long-styled form the elongated stigmas stand nearly upright, and alternate with the anthers. In this latter form the length of the stigmas varies considerably, their upper extremities projecting even a little above the anthers, or reaching up only to about their middle. Nevertheless, there is never the slightest difficulty in distinguishing between the two forms; for, besides the difference in the divergence of the stigmas, those of the short-styled form never reach even to the bases of the anthers. In this form the papillæ on the stigmatic surfaces are shorter, darker-coloured, and more crowded together than in the long-styled form; but these differences seem due merely to the shortening of the stigma, for in the varieties of the long-styled form with shorter stigmas, the papillæ are more crowded and darker-coloured than in those with the longer stigmas. Considering the slight and variable differences between the two forms of this Linum, it is not surprising that hitherto they have been overlooked.

In 1861 I had eleven plants in my garden, eight of which were long-styled, and three short-styled. Two very fine long-styled plants grew in a bed a hundred yards off all the others, and separated from them by a screen of evergreens. I marked twelve flowers, and placed on their stigmas a little pollen from the short-styled plants. The pollen of the two forms is, as stated, identical in appearance; the stigmas of the long-styled flowers were already thickly covered with their own pollen—so thickly that I could not find one bare stigma, and it was late in the season, namely, September 15th. Altogether, it seemed almost childish to expect any result. Nevertheless from my experiments on Primula, I had faith, and did not hesitate to make the trial, but certainly did not anticipate the full result which was obtained. The germens of these twelve flowers all swelled, and ultimately six fine capsules (the seed of which germinated on the following year) and two poor capsules were produced; only four capsules shanking off. These same two long-styled plants produced, in the course of the summer, a vast number of flowers, the stigmas of which were covered with their own pollen; but they all proved absolutely barren, and their germens did not even swell.

The nine other plants, six long-styled and three short-styled, grew not very far apart in my flower-garden. Four of these long-styled plants produced no seed-capsules; the fifth produced two; and the remaining one grew so close to a short-styled plant that their branches touched, and this produced twelve capsules, but they were poor ones. The case was different

with the short-styled plants.   The one which grew
close to the long-styled plant produced ninety-four
imperfectly fertilised capsules containing a multitude
of bad seeds, with a moderate number of good ones.
The two other short-styled plants growing together
were small, being partly smothered by other plants;
they did not stand very close to any long-styled plants,
yet they yielded together nineteen capsules.   These
facts seem to show that the short-styled plants are more
fertile with their own pollen than are the long-styled,
and we shall immediately see that this probably is the
case.   But I suspect that the difference in fertility between the two forms was in this instance in part due to
a distinct cause.   I repeatedly watched the flowers, and
only once saw a humble-bee momentarily alight on
one, and then fly away.   If bees had visited the several
plants, there cannot be a doubt that the four long-
styled plants, which did not produce a single capsule,
would have borne an abundance.   But several times I
saw small diptera sucking the flowers; and these
insects, though not visiting the flowers with anything
like the regularity of bees, would carry a little pollen
from one form to the other, especially when growing
near together; and the stigmas of the short-styled
plants, diverging within the tube of the corolla, would
be more likely than the upright stigmas of the long-
styled plants, to receive a small quantity of pollen if
brought to them by small insects.   Moreover from the
greater number of the long-styled than of the short-
styled plants in the garden, the latter would be more
likely to receive pollen from the long-styled, than the
long-styled from the short-styled.

In 1862 I raised thirty-four plants of this Linum in a
hot-bed; and these consisted of seventeen long-styled
and seventeen short-styled forms.   Seed sown later in the

flower-garden yielded seventeen long-styled and twelve short-styled forms. These facts justify the statement that the two forms are produced in about equal numbers. The thirty-four plants of the first lot were kept under a net which excluded all insects, except such minute ones as Thrips. I fertilised fourteen long-styled flowers legitimately with pollen from the short-styled, and got eleven fine seed-capsules, which contained on an average 8·6 seeds per capsule, but only 5·6 appeared to be good. It may be well to state that ten seeds is the maximum production for a capsule, and that our climate cannot be very favourable to this North-African plant. On three occasions the stigmas of nearly a hundred flowers were fertilised illegitimately with their own-form pollen, taken from separate plants, so as to prevent any possible ill effects from close inter-breeding. Many other flowers were also produced, which, as before stated, must have received plenty of their own pollen; yet from all these flowers, borne by the seventeen long-styled plants, only three capsules were produced. One of these included no seed, and the other two together gave only five good seeds. It is probable that this miserable product of two half-fertile capsules from the seventeen plants, each of which must have produced at least fifty or sixty flowers, resulted from their fertilisation with pollen from the short-styled plants by the aid of Thrips; for I made a great mistake in keeping the two forms under the same net, with their branches often interlocking; and it is surprising that a greater number of flowers were not accidentally fertilised.

Twelve short-styled flowers were in this instance castrated, and afterwards fertilised legitimately with pollen from the long-styled form; and they produced seven fine capsules. These included on an average

7·6 seeds, but of apparently good seed only 4·3 per
capsule. At three separate times nearly a hundred
flowers were fertilised illegitimately with their own-
form pollen, taken from separate plants; and numerous other flowers were produced, many of which
must have received their own pollen. From all these
flowers on the seventeen short-styled plants only fifteen
capsules were produced, of which only eleven contained any good seed, on an average 4·2 per capsule.
As remarked in the case of the long-styled plants,
some even of these capsules were perhaps the product
of a little pollen accidentally fallen from the adjoining
flowers of the other form on to the stigmas, or transported by Thrips. Nevertheless the short-styled plants
seem to be slightly more fertile with their own pollen
than the long-styled, in the proportion of fifteen capsules to three; nor can this difference be accounted
for by the short-styled stigmas being more liable to
receive their own pollen than the long-styled, for the
reverse is the case. The greater self-fertility of the
short-styled flowers was likewise shown in 1861 by
the plants in my flower-garden, which were left to
themselves, and were but sparingly visited by insects.

On account of the probability of some of the flowers
on the plants of both forms, which were covered under
the same net, having been legitimately fertilised in
an accidental manner, the relative fertility of the two
legitimate and two illegitimate unions cannot be
compared with certainty; but judging from the
number of good seeds per capsule, the difference was
at least in the ratio of 100 to 7, and probably much
greater.

Hildebrand tested my results, but only on a single
short-styled plant, by fertilising many flowers with
their own-form pollen; and these did not produce any

seed. This confirms my suspicion that some of the
few capsules produced by the foregoing seventeen
short-styled plants were the product of accidental
legitimate fertilisation. Other flowers on the same
plant were fertilised by Hildebrand with pollen from
the long-styled form, and all produced fruit.*

The absolute sterility (judging from the experiments of 1861) of the long-styled plants with their
own-form pollen led me to examine into its apparent
cause; and the results are so curious that they are
worth giving in detail. The experiments were tried
on plants grown in pots and brought successively
into the house.

First. Pollen from a short-styled plant was placed
on the five stigmas of a long-styled flower, and these,
after thirty hours, were found deeply penetrated by a
multitude of pollen-tubes, far too numerous to be
counted; the stigmas had also become discoloured
and twisted. I repeated this experiment on another
flower, and in eighteen hours the stigmas were penetrated by a multitude of long pollen-tubes. This is
what might have been expected, as the union is a
legitimate one. The converse experiment was likewise
tried, and pollen from a long-styled flower was placed
on the stigmas of a short-styled flower, and in twenty-
four hours the stigmas were discoloured, twisted, and
penetrated by numerous pollen-tubes; and this, again,
is what might have been expected, as the union was
a legitimate one.

Secondly. Pollen from a long-styled flower was placed
on all five stigmas of a long-styled flower on a separate
plant: after nineteen hours the stigmas were dissected,
and only a single pollen-grain had emitted a tube,

---

* 'Bot. Zeitung,' Jan. 1, 1864, p. 2.

and this was a very short one. To make sure that the pollen was good, I took in this case, and in most of the other cases, pollen either from the same anther or from the same flower, and proved it to be good by placing it on the stigma of a short-styled plant, and found numerous pollen-tubes emitted.

Thirdly. Repeated last experiment, and placed own-form pollen on all five stigmas of a long-styled flower; after nineteen hours and a half, not one single grain had emitted its tube.

Fourthly. Repeated the experiment, with the same result after twenty-four hours.

Fifthly. Repeated last experiment, and, after leaving pollen on for nineteen hours, put on an additional quantity of own-form pollen on all five stigmas. After an interval of three days, the stigmas were examined, and, instead of being discoloured and twisted, they were straight and fresh-coloured. Only one grain had emitted a quite short tube, which was drawn out of the stigmatic tissue without being ruptured.

The following experiments are more striking :—

Sixthly. I placed own-form pollen on three of the stigmas of a long-styled flower, and pollen from a short-styled flower on the other two stigmas. After twenty-two hours these two stigmas were discoloured, slightly twisted, and penetrated by the tubes of numerous pollen-grains: the other three stigmas, covered with their own-form pollen, were fresh, and all the pollen-grains were loose; but I did not dissect the whole stigma.

Seventhly. Experiment repeated in the same manner, with the same result.

Eighthly. Experiment repeated, but the stigmas were carefully examined after an interval of only five hours and a half. The two stigmas with pollen from a short-styled flower were penetrated by innumerable tubes, which were as yet short, and the stigmas themselves were not at all discoloured. The three stigmas covered with their own-form pollen were not penetrated by a single pollen-tube.

Ninthly. Put pollen of a short-styled flower on a single long-styled stigma, and own-form pollen on the other four stigmas; after twenty-four hours the one stigma was somewhat discoloured and twisted, and penetrated by many long tubes : the other four stigmas were quite straight and fresh ; but on dissecting them I found that three pollen-grains had protruded very short tubes into the tissue.

Tenthly. Repeated the experiment, with the same result after twenty-four hours, excepting that only two own-form grains had penetrated the stigmatic tissue with their tubes to a very short depth. The one stigma, which was deeply penetrated by a multitude of tubes from the short-styled pollen, presented a conspicuous difference in being much curled, half-shrivelled, and discoloured, in comparison with the other four straight and bright pink stigmas.

I could add other experiments; but those now given amply suffice to show that the pollen-grains of a short-styled flower placed on the stigma of a long-styled flower emit a multitude of tubes after an interval of from five to six hours, and penetrate the tissue ultimately to a great depth; and that after twenty-four hours the stigmas thus penetrated change colour, become twisted, and appear half-withered. On the other hand, pollen-grains from a long-styled flower placed on its own stigmas, do not emit their tubes after an interval of a day, or even three days; or at most only three or four grains out of a multitude emit their tubes, and these apparently never penetrate the stigmatic tissue deeply, and the stigmas themselves do not soon become discoloured and twisted.

This seems to me a remarkable physiological fact. The pollen-grains of the two forms are undistinguishable under the microscope; the stigmas differ only in length, degree of divergence, and in the size, shade of colour, and approximation of their papillæ, these latter differences being variable and apparently due merely to the degree of elongation of the stigma. Yet we plainly see that the two kinds of pollen and the two stigmas are widely dissimilar in their mutual reaction —the stigmas of each form being almost powerless on their own pollen, but causing, through some mysterious influence, apparently by simple contact (for I could detect no viscid secretion), the pollen-grains of the opposite form to protrude their tubes. It may be said that the two pollens and the two stigmas mutually recognise each other by some means. Taking fertility as the criterion of distinctness, it is no exaggeration to say that the pollen of the long-styled *Linum grandiflorum* (and conversely that of the other form) has been brought to a degree of differentiation, with respect to its action on the stigma of the same form, corresponding with that existing between the pollen and stigma of species belonging to distinct genera.

Linum perenne.—This species is conspicuously heterostyled, as has been noticed by several authors. The pistil in the long-styled form is nearly twice as long as that of the short-styled. In the latter the stigmas are smaller and, diverging to a greater degree, pass out low down between the filaments. I could detect no difference in the two forms in the size of the stigmatic papillæ. In the long-styled form alone the stigmatic surfaces of the mature pistils twist round, so as to face the circumference of the flower; but to this point I

shall presently return. Differently from what occurs in
L. grandiflorum, the long-styled flowers have stamens
hardly more than half the length of those in the short-
styled. The size of the pollen-grains is rather variable;
after some doubt, I have come to the conclusion that
there is no uniform difference between the grains in
the two forms. The long stamens in the short-styled
form project to some height above the corolla, and
their filaments are coloured blue apparently from exposure to the light. The anthers of the longer stamens
correspond in height with the lower part of the stigmas
of the long-styled flowers; and the anthers of the
shorter stamens of the latter correspond in the same
manner in height with the stigmas of the short-styled
flowers.

I raised from seed twenty-six plants, of which twelve
proved to be long-styled and fourteen short-styled.
They flowered well, but were not large plants. As I
did not expect them to flower so soon, I did not trans-
plant them, and they unfortunately grew with their
branches closely interlocked. All the plants were
covered under the same net, excepting one of each
form. Of the flowers on the long-styled plants, twelve
were illegitimately fertilised with their own-form pollen, taken in every case from a separate plant; and not
one set a seed-capsule: twelve other flowers were legitimately fertilised with pollen from short-styled flowers;
and they set nine capsules, each including on an
average 7 good seeds, ten being the maximum number
ever produced. Of the flowers on the short-styled
plants, twelve were illegitimately fertilised with own-
form pollen, and they yielded one capsule, including
only 3 good seeds; twelve other flowers were legitimately fertilised with pollen from long-styled flowers,
and these produced nine capsules, but one was bad;

the eight good capsules contained on an average 8
good seeds each.   Judging from the number of seeds
per capsule, the fertility of the two legitimate to that
of the two illegitimate unions is as 100 to 20.

The numerous flowers on the eleven long-styled
plants under the net, which were not fertilised, produced
only three capsules, including 8, 4, and 1 good seeds.
Whether these three capsules were the product of accidental legitimate fertilisation, owing to the branches
of the plants of the two forms interlocking, I will not
pretend to decide.   The single long-styled plant which
was left uncovered, and grew close by the uncovered
short-styled plant, produced five good pods; but it
was a poor and small plant.

The flowers borne on the thirteen short-styled plants
under the net, which were not fertilised, produced
twelve capsules, containing on an average 5·6 seeds.
As some of these capsules were very fine, and as five
were borne on one twig, I suspect that some minute
insect had accidentally got under the net and had
brought pollen from the other form to the flowers
which produced this little group of capsules.   The one
uncovered short-styled plant which grew close to the
uncovered long-styled plant yielded twelve capsules.

From these facts we have some reason to believe, as
in the case of *L. grandiflorum,* that the short-styled
plants are in a slight degree more fertile with their
own pollen than are the long-styled plants.   Anyhow
we have the clearest evidence, that the stigmas of each
form require for full fertility that pollen from the stamens of corresponding height belonging to the opposite
form should be brought to them.

Hildebrand, in the paper lately referred to, confirms
my results.   He placed a short-styled plant in his
house, and fertilised about 20 flowers with their own

pollen, and about 30 with pollen from another plant belonging to the same form, and these 50 flowers did not set a single capsule. On the other hand he fertilised about 30 flowers with pollen from the long-styled form, and these, with the exception of two, yielded capsules, containing good seeds.

It is a singular fact, in contrast with what occurred in the case of *L. grandiflorum*, that the pollen-grains of both forms of *L. perenne*, when placed on their own-form stigmas, emitted their tubes, though this action did not lead to the production of seeds. After an interval of eighteen hours, the tubes penetrated the stigmatic tissue, but to what depth I did not ascertain. In this case the impotence of the pollen-grains on their own stigmas must have been due either to the tubes not reaching the ovules, or to their not acting properly after reaching them.

The plants both of *L. perenne* and *grandiflorum* grew, as already stated, with their branches interlocked, and with scores of flowers of the two forms close together; they were covered by a rather coarse net, through which the wind, when high, passed; and such minute insects as Thrips could not, of course, be excluded; yet we have seen that the utmost possible amount of accidental fertilisation on seventeen long-styled plants in the one case, and on eleven long-styled plants in the other, resulted in the production, in each case, of three poor capsules; so that when the proper insects are excluded, the wind does hardly anything in the way of carrying pollen from plant to plant. I allude to this fact because botanists, in speaking of the fertilisation of various flowers, often refer to the wind or to insects as if the alternative were indifferent. This view, according to my experience, is entirely erroneous. When the wind is the agent in carrying pollen, either from

one sex to the other, or from hermaphrodite to hermaphrodite, we can recognise structure as manifestly adapted to its action as to that of insects when these are
the carriers. We see adaptation to the wind in the in-
coherence of the pollen,—in the inordinate quantity
produced (as in the Coniferæ, Spinage, &c.),—in the
dangling anthers well fitted to shake out the pollen,—
in the absence or small size of the perianth,—in the
protrusion of the stigmas at the period of fertilisation,
—in the flowers being produced before they are hidden
by the leaves,—and in the stigmas being downy or
plumose (as in the Gramineæ, Docks, &c.), so as to
secure the chance-blown grains. In plants which are
fertilised by the wind, the flowers do not secrete nectar,
their pollen is too incoherent to be easily collected by
insects, they have not bright-coloured corollas to serve
as guides, and they are not, as far as I have seen,
visited by insects. When insects are the agents of fertilisation (and this is incomparably the more frequent
case with hermaphrodite plants), the wind plays no
part, but we see an endless number of adaptations to
ensure the safe transport of the pollen by the living
workers. These adaptations are most easily recognised
in irregular flowers; but they are present in regular
flowers, of which those of Linum offer a good instance,
as I will now endeavour to show.

I have already alluded to the rotation of each separate stigma in the long-styled form of *Linum perenne*.
In both forms of the other heterostyled species and in
the homostyled species of Linum which I have seen,
the stigmatic surfaces face the centre of the flower,
with the furrowed backs of the stigmas, to which the
styles are attached, facing outwards. This is the case
with the stigmas of the long-styled flowers of *L.
perenne* whilst in bud. But by the time the flowers

have expanded, the five stigmas twist round so as to face the circumference, owing to the torsion of that part of the style which lies beneath the stigma. I should state that the five stigmas do not always turn round completely, two or three sometimes facing only obliquely outwards. My observations were made during October; and it is not improbable that earlier in the season the torsion would have been more complete; for after two or three cold and wet days the movement was very imperfectly performed. The flowers should be examined shortly after their expansion, as their duration is brief; as soon as they begin to wither, the styles become spirally twisted all together, the original position of the parts being thus lost.

He who will compare the structure of the whole flower in both forms of *L. perenne* and *grandiflorum*, and, as I may add, of *L. flavum*, will not doubt about the meaning of this torsion of the styles in the one form alone of *L. perenne*, as well as the meaning of the divergence of the stigmas in the short-styled form of all three species. It is absolutely necessary as we know, that insects should carry pollen from the flowers of the one form reciprocally to those of the other. Insects are attracted by five drops of nectar, secreted exteriorly at the base of the stamens, so that to reach these drops they must insert their proboscides outside the ring of broad filaments, between them and the petals. In the short-styled form of the above three species, the stigmas face the axis of the flower; and had the styles retained their original upright and central position, not only would the stigmas have presented their backs to the insects which sucked the flowers, but their front and fertile surfaces would have been separated from the entering insects by the ring of broad filaments, and would never have received any pollen. As it is, the styles diverge and pass out between the filaments. After this movement the short stigmas lie within the tube of the corolla; and their papillous surfaces being now turned upwards are necessarily brushed by every entering insect, and thus receive the required pollen.

In the long-styled form of *L. grandiflorum*, the almost parallel or slightly diverging anthers and stigmas project a little above the tube of the somewhat concave flower; and they stand directly over the open space leading to the drops of nectar. Consequently when insects visit the flowers of either form (for the stamens in this species occupy the same position in both forms), they will get their foreheads or proboscides well dusted with the coherent pollen. As soon as they visit the flowers of the long-styled form they will necessarily leave pollen on the proper surface of the elongated stigmas; and when they visit the short-styled flowers, they will leave pollen on the upturned stigmatic surfaces. Thus the stigmas of both forms will receive indifferently the pollen of both forms; but we know that the pollen alone of the opposite form causes fertilisation.

In the case of *L. perenne*, affairs are arranged more perfectly; for the stamens in the two forms stand at different heights, so that pollen from the anthers of the longer stamens will adhere to one part of an insect's body, and will afterwards be brushed off by the rough stigmas of the longer pistils; whilst pollen from the anthers of the shorter stamens will adhere to a different part of the insect's body, and will afterwards be brushed off by the stigmas of the shorter pistils; and this is what is required for the legitimate fertilisation of both forms. The corolla of *L. perenne* is more expanded than that of *L. grandiflorum*, and the stigmas of the long-styled form do not diverge greatly from one another; nor do the stamens of either form. Hence insects, especially rather small ones, will not insert their proboscides between the stigmas of the long-styled form, nor between the anthers of either form (Fig. 5), but will strike against them, at nearly right angles, with the backs of their head or thorax. Now, in the long-styled flowers, if each stigma did

Fig. 5.

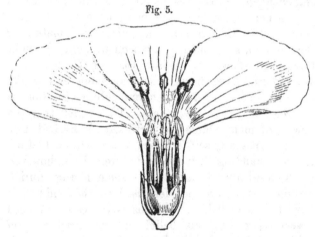

Long-styled form of L. PERENNE, var. *Austriacum* in its early condition before the stigmas have rotated. The petals and calyx have been removed on the near side.*

not rotate on its axis, insects in visiting them would strike their heads against the backs of the stigmas; as it is, they strike against that surface which is covered

* I neglected to get drawings made from fresh flowers of the two forms. But Mr. Fitch has made the above sketch of a long-styled flower from dried specimens and from published engravings. His well-known skill ensures accuracy in the proportional size of the parts.

H

with papillæ, with their heads already charged with pollen from the stamens of corresponding height borne by the flowers of the other form, and legitimate fertilisation is thus ensured.

Thus we can understand the meaning of the torsion of the styles in the long-styled flowers alone, as well as their divergence in the short-styled flowers.

One other point is worth notice. In botanical works many flowers are said to be fertilised in the bud. This statement generally rests, as far as I can discover, on the anthers opening in the bud; no evidence being adduced that the stigma is at this period mature, or that it is not subsequently acted on by pollen brought from other flowers. In the case of *Cephalanthera grandiflora* I have shown[*] that precocious and partial self-fertilisation, with subsequent full fertilisation, is the regular course of events. The belief that the flowers of many plants are fertilised in the bud, that is, are perpetually self-fertilised, is a most effectual bar to understanding their real structure. I am, however, far from wishing to assert that some flowers, during certain seasons, are not fertilised in the bud; for I have reason to believe that this is the case. A good observer,[†] resting his belief on the usual kind of evidence, states that in *Linum Austriacum* (which is heterostyled, and is considered by Planchon as a variety of *L. perenne*) the anthers open the evening before the expansion of the flowers, and that the stigmas are then almost always fertilised. Now we know positively that, so far from *Linum perenne* being fertilised by its own pollen in the bud, its own pollen is as powerless on the stigma as so much inorganic dust.

Linum flavum.—The pistil of the long-styled form

---

[*] 'Fertilisation of Orchids,' p. 108.—2nd edit. 1877, p. 84.

[†] 'Etudes sur la Géogr. Bot.,' H. Lecoq, 1856, tom. v. p. 325.

of this species is nearly twice as long as that of
the short-styled; the stigmas are longer and the
papillæ coarser. In the short-styled form the stigmas
diverge and pass out between the filaments, as in the
previous species. The stamens in the two forms differ
in length; and, what is singular, the anthers of the
longer stamens are not so long as those of the other
form; so that in the short-styled form both the stigmas
and the anthers are shorter than in the long-styled
form. The pollen-grains of the two forms do not differ
in size. As this species is propagated by cuttings,
generally all the plants in the same garden belong to
the same form. I have inquired, but have never heard
of its seeding in this country. Certainly my own plants
never produced a single seed as long as I possessed
only one of the two forms. After considerable search
I procured both forms, but from want of time only a few
experiments were made. Two plants of the two forms
were planted some way apart in my garden, and were
not covered by nets. Three flowers on the long-styled
plant were legitimately fertilised with pollen from the
short-styled plant, and one of them set a fine capsule.
No other capsules were produced by this plant. Three
flowers on the short-styled plant were legitimately
fertilised with pollen from the long-styled, and all
three produced capsules, containing respectively no
less than 8, 9, and 10 seeds. Three other flowers on
this plant, which had not been artificially fertilised,
produced capsules containing 5, 1, and 5 seeds; and
it is quite possible that pollen may have been
brought to them by insects from the long-styled plant
growing in the same garden. Nevertheless, as they
did not yield half the number of seeds compared
with the other flowers on the same plant which had
been artificially and legitimately fertilised, and as the

H 2

short-styled plants of the two previous species apparently evince some slight capacity for fertilisation with their own-form pollen, these three capsules may have been the product of self-fertilisation.

Besides the three species now described, the yellow-flowered *L. corymbiferum* is certainly heterostyled, as is, according to Planchon,* *L. salsoloides.* This botanist is the only one who seems to have inferred that heterostylism might have some important functional bearing. Dr. Alefeld, who has made a special study of the genus, says† that about half of the sixty-five species known to him are heterostyled. This is the case with *L. trigynum*, which differs so much from the other species that it has been formed by him into a distinct genus.‡ According to the same author, none of the species which inhabit America and the Cape of Good Hope are heterostyled.

I have examined only three homostyled species, namely, *L. usitatissimum, angustifolium,* and *catharticum.* I raised 111 plants of a variety of the first-named species, and these, when protected under a net, all produced plenty of seed. The flowers, according to H. Müller,§ are frequented by bees and moths. With respect to *L. catharticum,* the same author shows that the flowers are so constructed that they can freely fertilise themselves; but if visited by insects they might be cross-fertilised. He has, however, only once seen the flowers thus visited during the day; but it

---

* Hooker's 'London Journal of Botany,' 1848, vol. vii. p. 174.

† 'Bot. Zeitung,' Sep. 18th, 1863, p. 281.

‡ It is not improbable that the allied genus, Hugonia, is heterostyled, for one species is said by Planchon (Hooker's 'London Journal of Botany,' 1848, vol. vii. p. 525) to be provided with "staminibus exsertis;" another with "stylis staminibus longioribus," and another has "stamina 5, majora, stylos longe superantia."

§ 'Die Befruchtung der Blumen,' &c., p. 168.

may be suspected that they are frequented during the night by small moths for the sake of the five minute drops of nectar secreted. Lastly, *L. Lewisii* is said by Planchon to bear on the same plant flowers with stamens and pistils of the same height, and others with the pistils either longer or shorter than the stamens. This case formerly appeared to me an extraordinary one; but I am now inclined to believe that it is one merely of great variability.[*]

PULMONARIA (BORAGINEÆ).

Pulmonaria officinalis.—Hildebrand has published [†] a full account of this heterostyled plant. The pistil of the long-styled form is twice as long as that of the short-styled; and the stamens differ in a corresponding, though converse, manner. There is no marked difference in the shape or state of surface of the stigma in the two forms. The pollen-grains of the short-styled form are to those of the long-styled as 9 to 7, or as 100 to 78, in length, and as 7 to 6 in breadth. They do not differ in the appearance of their contents. The corolla of the one form differs in shape from that of the other in nearly the same manner as in Primula; but besides this difference the flowers of the short-styled are generally the larger of the two. Hildebrand collected on the Siebengebirge, ten wild long-styled and ten short-styled plants. The former bore 289 flowers, of which 186 (i.e. 64 per cent.) had set fruit, yielding 1·88 seed per fruit. The ten short-styled plants bore 373 flowers, of which 262 (i.e.

---

[*] Planchon, in Hooker's 'London Journal of Botany,' 1848, vol. vii. p. 175. See on this subject Asa Gray, in 'American Journal of Science,' vol. xxxvi. Sept. 1863, p. 284.

[†] 'Bot Zeitung,' 1865, Jan. 13, p. 13.

70 per cent.) had set fruit, yielding 1·86 seed per fruit. So that the short-styled plants produced many more flowers, and these set a rather larger proportion of fruit, but the fruits themselves yielded a slightly lower average number of seeds than did the long-styled plants. The results of Hildebrand's experiments on the fertility of the two forms are given in the following table :—

TABLE 19.

Pulmonaria officinalis (from Hildebrand).

Nature of Union.	Number of Flowers fertilised.	Number of Fruits produced.	Average Number of Seeds per Fruit.
Long-styled flowers, by pollen of short-styled. Legitimate union	14	10	1·30
Long-styled flowers, 14 by own-pollen, and 16 by pollen of other plant of same form. Illegitimate union	30	0	0
Short-styled flowers, by pollen of long-styled. Legitimate union	16	14	1·57
Short-styled flowers, 11 by own pollen, 14 by pollen of other plant of same form. Illegitimate union	25	0	0

In the summer of 1864, before I had heard of Hildebrand's experiments, I noticed some long-styled plants of this species (named for me by Dr. Hooker) growing by themselves in a garden in Surrey; and to my surprise about half the flowers had set fruit, several of which contained 2, and one contained even 3 seeds. These seeds were sown in my garden and eleven seedlings thus raised, all of which proved long-styled, in accordance with the usual rule in such cases. Two years afterwards the plants were left uncovered, no other plant of the same genus growing in my garden, and the flowers were visited by many bees. They set an abundance of seeds : for instance, I gathered from a single plant rather less than half of the seeds which it had produced, and they numbered 47. Therefore this illegitimately fertilised plant must have produced about 100 seeds ; that is, thrice as many as one of the wild long-styled plants collected on the Siebengebirge by Hildebrand, and which, no doubt, had been legitimately fertilised. In the following year one of my plants was covered by a net, and even under these unfavourable conditions it produced spontaneously a few seeds. It should be observed that as the flowers stand either almost horizontally or hang considerably downwards, pollen from the short stamens would be likely to fall on the stigma. We thus see that the English long-styled plants when illegitimately fertilised were highly fertile, whilst the German plants similarly treated by Hildebrand were completely sterile. How to account for this wide discordance in our results I know not. Hildebrand cultivated his plants in pots and kept them for a time in the house, whilst mine were grown out of doors ; and he thinks that this difference of treatment may have caused the difference in our results. But this does not appear to me nearly a sufficient cause, although his plants were slightly less productive than the wild ones growing on the Siebengebirge. My plants exhibited no tendency to become equal-styled, so as to lose their proper long-styled character, as not rarely happens under cultivation with several heterostyled species of Primula ; but it would appear that they had been greatly affected in function, either by long-continued cultivation or by some other cause. We shall see in a future chapter that heterostyled plants illegitimately

fertilised during several successive generations sometimes become more self-fertile ; and this may have
been the case with my stock of the present species
of Pulmonaria ; but in this case we must assume
that the long-styled plants were at first sufficiently
fertile to yield some seed, instead of being absolutely
self-sterile like the German plants.

Pulmonaria angustifolia.—Seedlings of this plant,
raised from plants growing wild in the Isle of Wight,
were named for me by Dr. Hooker. It is so closely
allied to the last species, differing chiefly in the shape
and spotting of the leaves, that the two have been considered by several eminent botanists—for instance,
Bentham—as mere varieties. But, as we shall presently
see, good evidence can be assigned for ranking them
as distinct. Owing to the doubts on this head, I tried
whether the two would mutually fertilise one another.
Twelve short-styled flowers of *P. angustifolia* were
legitimately fertilised with pollen from long-styled
plants of *P. officinalis* (which, as we have just seen, are
moderately self-fertile), but they did not produce a
single fruit. Thirty-six long-styled flowers of *P.
angustifolia* were also illegitimately fertilised during
two seasons with pollen from the long-styled *P.
officinalis,* but all these flowers dropped off unimpregnated. Had the plants been mere varieties of
the same species these illegitimate crosses would
probably have yielded some seeds, judging from my
success in illegitimately fertilising the long-styled
flowers of *P. officinalis ;* and the twelve legitimate
crosses, instead of yielding no fruit, would almost
certainly have yielded a considerable number, namely,
about nine, judging from the results given in the following table (20). Therefore *P. officinalis* and *angustifolia* appear to be good and distinct species, in

conformity with other important functional differences between them, immediately to be described.

The long-styled and short-styled flowers of *P. angustifolia* differ from one another in structure in nearly the same manner as those of *P. officinalis*. But in the accompanying figure a slight bulging of the corolla

Fig. 6.

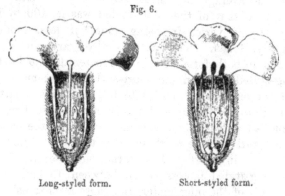

Long-styled form.   Short-styled form.

PULMONARIA ANGUSTIFOLIA.

in the long-styled form, where the anthers are seated, has been overlooked. My son William, who examined a large number of wild plants in the Isle of Wight, observed that the corolla, though variable in size, was generally larger in the long-styled flowers than in the short-styled; and certainly the largest corollas of all were found on the long-styled plants, and the smallest on the short-styled. Exactly the reverse occurs, according to Hildebrand, with *P. officinalis*. Both the pistils and stamens of *P. angustifolia* vary much in length; so that in the short-styled form the distance between the stigma and the anthers varied from 119 to 65 divisions of the micrometer, and in the long-styled from 115 to 112. From an average of seven measurements of each form the distance between these organs in the long-styled is to the same distance in the short-styled form as 100 to 69 ; so that the stigma in the one form does not stand on a level with the anthers in the other. The long-styled pistil is some-times thrice as long as that of the short-styled ; but from an average of ten measurements of both, its length to that of the short-styled was as 100 to 56. The stigma varies in being more or less, though slightly, lobed. The anthers also vary much in length in both forms, but in a greater degree in the long-styled than in the short-styled form ; many in the former being from 80 to 63, and in the latter from 80 to 70 divisions of the micrometer in length. From an average of seven measurements, the short-styled anthers were to those from the long-styled as 100 to 91 in length. Lastly, the pollen-grains from the long-styled flowers varied between 13 and 11·5 divisions of the micrometer, and those from the short-styled between 15 and 13. The average diameter of 25 grains from the latter, or short-styled form, was to that of 20 grains from the long-styled as 100 to 91. We see, therefore, that the pollen-grains from the smaller anthers of the shorter stamens in the long-styled form are, as usual, of smaller size than those in the other form. But what is remarkable, a larger proportion of the grains were small, shrivelled, and worthless. This could be seen by merely comparing the contents of the anthers from several distinct plants of each form. But in one instance my son found, by counting, that out of 193 grains from a long-styled flower, 53 were bad, or 27 per cent. ; whilst out of 265 grains from a short-styled flower only 18 were bad, or 7 per cent. From the condition of the pollen in the long-styled form, and from the extreme variability of all the organs in both forms, we may perhaps suspect that the plant is undergoing a change, and tending to become diœcious.

My son collected in the Isle of Wight on two occasions 202 plants, of which 125 were long-styled and 77 short-styled ; so that the former were the more numerous.   On the other hand, out of 18 plants raised by me from seed, only 4 were long-styled and 14 short-styled.   The short-styled plants seemed to my son to produce a greater number of flowers than the long-styled; and he came to this conclusion before a similar statement had been published by Hildebrand with respect to *P. officinalis.*   My son gathered ten branches from ten different plants of both forms, and found the number of flowers of the two forms to be as 100 to 89, 190 being short-styled and 169 long-styled. With *P. officinalis* the difference, according to Hildebrand, is even greater, namely, as 100 flowers for the short-styled to 77 for the long-styled plants.   The following table shows the results of my experiments :—

TABLE 20.

Pulmonaria angustifolia.

Nature of the Union.	Number of Flowers fertilised.	Number of Fruits produced.	Average Number of Seeds per Fruit.
Long-styled flowers, by pollen of short-styled. Legitimate union	18	9	2·11
Long-styled flowers, by own-form pollen. Illegitimate union	18	0	0
Short-styled flowers, by pollen of long-styled. Legitimate union	18	15	2·60
Short-styled flowers, by own-form pollen. Illegitimate union	12	7	1·86

We see in this table that the fertility of the two legitimate unions to that of the two illegitimate together is as 100 to 35, judged by the proportion of flowers which produced fruit; and as 100 to 32, judged by the average number of seeds per fruit. But the small number of fruit yielded by the 18 long-styled flowers in the first line was probably accidental, and if so, the difference in the proportion of legitimately and illegitimately fertilised flowers which yield fruit is really greater than that represented by the ratio of 100 to 35. The 18 long-styled flowers illegitimately fertilised yielded no seeds,—not even a vestige of one. Two long-styled plants which were placed under a net produced 138 flowers, besides those which were artificially fertilised, and none of these set any fruit; nor did some plants of the same form which were protected during the next summer. Two other long-styled plants were left uncovered (all the short-styled plants having been previously covered up), and humble-bees, which had their foreheads white with pollen, incessantly visited the flowers, so that their stigmas must have received an abundance of pollen, yet these flowers did not produce a single fruit. We may therefore conclude that the long-styled plants are absolutely barren with their own-form pollen, though brought from a distinct plant. In this respect they differ greatly from the long-styled English plants of *P. officinalis* which were found by me to be moderately self-fertile; but they agree in their behaviour with the German plants of *P. officinalis* experimented on by Hildebrand.

Eighteen short-styled flowers legitimately fertilised yielded, as may be seen in Table 20, 15 fruits, each having on an average 2·6 seeds. Four of these fruits contained the highest possible number of seeds, namely

4, and four other fruits contained each 3 seeds. The 12 illegitimately fertilised short-styled flowers yielded 7 fruits, including on an average 1·86 seed; and one of these fruits contained the maximum number of 4 seeds. This result is very surprising in contrast with the absolute barrenness of the long-styled flowers when illegitimately fertilised; and I was thus led to attend carefully to the degree of self-fertility of the short-styled plants. A plant belonging to this form and covered by a net bore 28 flowers besides those which had been artificially fertilised, and of all these only two produced a fruit each including a single seed. This high degree of self-sterility no doubt depended merely on the stigmas not receiving any pollen, or not a sufficient quantity. For after carefully covering all the long-styled plants in my garden, several short-styled plants were left exposed to the visits of humble-bees, and their stigmas will thus have received plenty of short-styled pollen; and now about half the flowers, thus illegitimately fertilised, set fruit. I judge of this proportion partly from estimation and partly from having examined three large branches, which had borne 31 flowers, and these produced 16 fruits. Of the fruits produced 233 were collected (many being left ungathered), and these included on an average 1·82 seed. No less than 16 out of the 233 fruits included the highest possible number of seeds, namely 4, and 31 included 3 seeds. So we see how highly fertile these short-styled plants were when illegitimately fertilised with their own-form pollen by the aid of bees.

The great difference in the fertility of the long and short-styled flowers, when both are illegitimately fertilised, is a unique case, as far as I have observed with heterostyled plants. The long-styled flowers when thus fertilised are utterly barren, whilst about half of the short-styled ones produce capsules, and these include a little above two-thirds of the number of seeds yielded by them when legitimately fertilised. The sterility of the illegitimately fertilised long-styled flowers is probably increased by the deteriorated condition of their pollen; nevertheless this pollen was highly efficient when applied to the stigmas of the short-styled flowers. With several species of Primula the short-styled flowers are much more sterile than the long-styled, when both are illegitimately fertilised; and it is a tempting view, as formerly remarked, that this greater sterility of the short-styled flowers is a special adaptation to check self-fertilisation, as their stigmas are eminently liable to receive their own pollen. This view is even still more tempting in the case of the long-styled form of *Linum grandiflorum.* On the other hand, with *Pulmonaria angustifolia,* it is evident, from the corolla projecting obliquely upwards, that pollen is much more likely to fall on, or to be carried by insects down to the stigma of the short-styled than of the long-styled flowers; yet the short-styled instead of being more sterile, as a protection against self-fertilisation, are far more fertile than the long-styled, when both are illegitimately fertilised.

Pulmonaria azurea, according to Hildebrand, is not heterostyled.[*]

From an examination of dried flowers of *Amsinckia spectabilis,* sent me by Prof. Asa Gray, I formerly thought that this plant, a member of the Boragineæ, was heterostyled. The pistil varies to an extraordinary degree in length, being in some specimens twice as long as in others, and the point of insertion of the stamens likewise varies. But on raising many plants from seed, I soon became convinced that the whole case was one of mere variability. The first-formed flowers are apt to

---

[*] 'Die Geschlechter-Vertheilung bei den Pflanzen,' 1867, p. 37.

have stamens somewhat arrested in development, with very little pollen in their anthers; and in such flowers the stigma projects above the anthers, whilst generally it stands below and sometimes on a level with them. I could detect no difference in the size of the pollen-grain or in the structure of the stigma in the plants which differed most in the above respects; and all of them, when protected from the access of insects, yielded plenty of seeds. Again, from statements made by Vaucher, and from a hasty inspection, I thought at first that the allied *Anchusa arvensis* and *Echium vulgae* were heterostyled, but soon saw my error. From information given me, I examined dried flowers of another member of the Boragineæ, *Arnebia hispidissima*, collected from several sites, and though the corolla, together with the included organs, differed much in length, there was no sign of heterostylism.

POLYGONUM FAGOPYRUM (POLYGONACEÆ).

Hildebrand has shown that this plant, the common Buck-wheat, is heterostyled.[*] In the long-styled form (Fig. 7), the three stigmas project considerably above the eight short stamens, and stand on a level with the anthers of the eight long stamens in the short-styled form; and so it is conversely with the stigmas and stamens of this latter form. I could perceive no difference in the structure of the stigmas in the two forms. The pollen-grains of the short-styled form are to those of the long-styled as 100 to 82 in diameter. This plant is therefore without doubt heterostyled.

I experimented only in an imperfect manner on the relative fertility of the two forms. Short-styled flowers were dragged several times over two heads of flowers on long-styled plants, protected under a net, which were thus legitimately, though not fully, fertilised. They produced 22 seeds, or 11 per flower-head.

Three flower-heads on long-styled plants received

---

[*] 'Die Geschlechter-Vertheilung,' &c., 1867, p. 34.

pollen in the same manner from other long-styled plants, and were thus illegitimately fertilised. They produced 14 seeds, or only 4·66 per flower-head.

Two flower-heads on short-styled plants received pollen in like manner from long-styled flowers, and were thus legitimately fertilised. They produced 8 seeds, or 4 per flower-head.

Fig. 7.

Upper figure, the long-styled form; lower figure, the short-styled. Some of the anthers have dehisced, others have not.

POLYGONUM FAGOPYRUM.   (From H. Müller.)

Four heads on short-styled plants similarly received pollen from other short-styled plants, and were thus illegitimately fertilised. They produced 9 seeds, or 2·25 per flower-head.

The results from fertilising the flower-heads in the above imperfect manner cannot be fully trusted; but I may state that the four legitimately fertilised flower-heads yielded on an average 7·50 seeds per head; whereas the seven illegitimately fertilised heads yielded less than half the number, or on an average only 3·28 seeds. The legitimately crossed seeds from the long-styled flowers were finer than those from the illegitimately fertilised flowers on the same plants, in the ratio of 100 to 82, as shown by the weights of an equal number.

About a dozen plants, including both forms, were protected under nets, and early in the season they produced spontaneously hardly any seeds, though at this period the artificially fertilised flowers produced an abundance; but it is a remarkable fact that later in the season, during September, both forms became highly self-fertile. They did not, however, produce so many seeds as some neighbouring uncovered plants which were visited by insects. Therefore the flowers of neither form when left to fertilise themselves late in the season without the aid of insects, are nearly so sterile as most other heterostyled plants. A large number of insects, namely 41 kinds as observed by H. Müller,* visit the flowers for the sake of the eight drops of nectar. He infers from the structure of the flowers that insects would be apt to fertilise them both illegitimately as well as legitimately; but he is mistaken in supposing that the long-styled flowers cannot spontaneously fertilise themselves.

Differently to what occurs in the other genera hitherto noticed, Polygonum, though a very large genus, contains, as far as is at present known, only a single heterostyled species, namely the present one. H. Müller in his interesting description of several

---

* ' Die Befruchtung,' &c., p. 175, and ' Nature,' Jan. 1, 1874, p. 166.

other species shows that *P. bistorta* is so strongly proterandrous (the anthers generally falling off before the
stigmas are mature) that the flowers must be cross-
fertilised by the many insects which visit them. Other
species bear much less conspicuous flowers which secrete little or no nectar, and consequently are rarely
visited by insects; these are adapted for self-fertilisation, though still capable of cross-fertilisation. According to Delpino, the Polygonaceæ are generally
fertilised by the wind, instead of by insects as in the
present genus.

LEUCOSMIA BURNETTIANA (THYMELIÆ).

As Prof. Asa Gray has expressed his belief [*] that this species
and *L. acuminata*, as well as some species in the allied genus
Drymispermum, are dimorphic or heterostyled, I procured
from Kew, through the kindness of Dr. Hooker, two dried
flowers of the former species, an inhabitant of the Friendly
Islands in the Pacific. The pistil of the long-styled form is to
that of the short-styled as 100 to 86 in length; the stigma,
projects just above the throat of the corolla, and is surrounded
by five anthers, the tips of which reach up almost to its base;
and lower down, within the tubular corolla, five other and
rather smaller anthers are seated. In the short-styled form,
the stigma stands some way down the tube of the corolla, nearly
on a level with the lower anthers of the other form: it differs
remarkably from the stigma of the long-styled form, in being
more papillose, and in being longer in the ratio of 100 to 60.
The anthers of the upper stamens in the short-styled form are
supported on free filaments, and project above the throat of the
corolla, whilst the anthers of the lower stamens are seated in
the throat on a level with the upper stamens of the other form.
The diameters of a considerable number of grains from both sets
of anthers in both forms were measured, but they did not differ
in any trustworthy degree. The mean diameter of twenty-two

---

[*] 'American Journal of Science,' 1865, p. 101, and Seemann's
'Journal of Botany,' vol. iii. 1865,
p. 305.

grains from the short-styled flower was to that of twenty-four grains from the long-styled, as 100 to 99. The anthers of the upper stamens in the short-styled form appeared to be poorly developed, and contained a considerable number of shrivelled grains which were omitted in striking the above average. Notwithstanding the fact of the pollen-grains from the two forms not differing in diameter in any appreciable degree, there can hardly be a doubt from the great difference in the two forms in the length of the pistil, and especially of the stigma, together with its more papillose condition in the short-styled form, that the present species is truly heterostyled. This case resembles that of *Linum grandiflorum*, in which the sole difference between the two forms consists in the length of the pistils and stigmas. From the great length of the tubular corolla of Leucosmia, it is clear that the flowers are cross-fertilised by large Lepidoptera or by honey-sucking birds, and the position of the stamens in two whorls one beneath the other, which is a character that I have not seen in any other heterostyled dimorphic plant, probably serves to smear the inserted organ thoroughly with pollen.

MENYANTHES TRIFOLIATA (GENTIANEÆ).

This plant inhabits marshes: my son William gathered 247 flowers from so many distinct plants, and of these 110 were long-styled, and 137 short-styled. The pistil of the long-styled form is in length to that of the short-styled in the ratio of about 3 to 2. The stigma of the former, as my son observed, is decidedly larger than that of the short-styled; but in both forms it varies much in size. The stamens of the short-styled are almost double the length of those of the long-styled; so that their anthers stand rather above the level of the stigma of the long-styled form. The anthers also vary much in size, but seem often to be of larger size in the short-styled flowers. My son made with the camera many drawings of the pollen-grains, and those from the short-styled flowers were in diameter in nearly the ratio of 100 to 84 to those from the long-styled flowers. I know nothing about the capacity for fertilisation in the two forms; but short-styled plants, living by themselves in the gardens at Kew, have produced an abundance of capsules, yet the seeds have never germinated; and this looks as if the short-styled form was sterile with its own pollen.

I 2

LIMNANTHEMUM INDICUM (GENTIANEÆ).

This plant is mentioned by Mr. Thwaites in his Enumeration of the Plants of Ceylon as presenting two forms; and he was so kind as to send me specimens preserved in spirits. The pistil of the long-styled form is nearly thrice as long (i.e. as 14 to 5) as that of the short-styled, and is very much thinner in the ratio of about 3 to 5. The foliaceous stigma is more expanded, and twice as large as that of the short-styled form. In the latter the stamens are about twice as long as those of the long-styled, and their anthers are larger in the ratio of 100 to 70. The pollen-grains, after having been long kept in spirits, were of the same shape and size in both forms. The ovules, according to Mr. Thwaites, are equally numerous (viz. from 70 to 80) in the two forms.

VILLARSIA [SP. ?] (GENTIANEÆ).

Fritz Müller sent me from South Brazil dried flowers of this aquatic plant, which is closely allied to Limnanthemum. In the long-styled form the stigma stands some way above the anthers, and the whole pistil, together with the ovary, is in length to that of the short-styled form as about 3 to 2. In the latter form the anthers stand above the stigma, and the style is very short and thick; but the pistil varies a good deal in length, the stigma being either on a level with the tips of the sepals or considerably beneath them. The foliaceous stigma in the long-styled form is larger, with the expansions running farther down the style, than in the other form. One of the most remarkable differences between the two forms is that the anthers of the longer stamens in the short-styled flowers are conspicuously longer than those of the shorter stamens in the long-styled flowers. In the former the sub-triangular pollen-grains are larger; the ratio between their breadth (measured from one angle to the middle of the opposite side) and that of the grains from the long-styled flowers being about 100 to 75. Fritz Müller also informs me that the pollen of the short-styled flowers has a bluish tint, whilst that of the long-styled is yellow. When we treat of *Lythrum salicaria* we shall find a strongly marked contrast in the colour of the pollen in two of the forms.

The three genera, Menyanthes, Limnanthemum, and Villarsia, now described, constitute a well-marked sub-tribe of the Gentianeæ. All the species, as far as at present known, are heterostyled, and all inhabit aquatic or sub-aquatic stations.

FORSYTHIA SUSPENSA (OLEACEÆ).

Professor Asa Gray states that the plants of this species growing in the Botanic Gardens at Cambridge, U.S., are short-styled,
but that Siebold and Zuccarini describe the long-styled form,
and give figures of two forms; so that there can be little doubt,
as he remarks, about the plant being dimorphic.* I therefore
applied to Dr. Hooker, who sent me a dried flower from Japan,
another from China, and another from the Botanic Gardens at
Kew. The first proved to be long-styled, and the other two
short-styled. In the long-styled form, the pistil is in length
to that of the short-styled as 100 to 38, the lobes of the stigma
being a little longer (as 10 to 9), but narrower and less divergent. This last character, however, may be only a temporary
one. There seems to be no difference in the papillose condition
of the two stigmas. In the short-styled form, the stamens are
in length to those of the long-styled as 100 to 66, but the anthers
are shorter in the ratio of 87 to 100; and this is unusual, for
when there is any difference in size between the anthers of the
two forms, those from the longer stamens of the short-styled are
generally the longest. The pollen-grains from the short-styled
flowers are certainly larger, but only in a slight degree, than
those from the long-styled, namely, as 100 to 94 in diameter.
The short-styled form, which grows in the Gardens at Kew, has
never there produced fruit.

Forsythia viridissima appears likewise to be heterostyled; for
Professor Asa Gray says that although the long-styled form
alone grows in the gardens at Cambridge, U.S., the published
figures of this species belong to the short-styled form.

CORDIA [SP. ?] (CORDIACEÆ).

Fritz Müller sent me dried specimens of this shrub, which he
believes to be heterostyled; and I have not much doubt that
this is the case, though the usual characteristic differences are
not well pronounced in the two forms. *Linum grandiflorum*
shows us that a plant may be heterostyled in function in the
highest degree, and yet the two forms may have stamens of
equal length, and pollen-grains of equal size. In the present
species of Cordia, the stamens of both forms are of nearly equal

---

* 'The American Naturalist,' July 1873, p. 422.

length, those of the short-styled being rather the longest; and
the anthers of both are seated in the mouth of the corolla. Nor
could I detect any difference in the size of the pollen-grains,
when dry or after being soaked in water. The stigmas of the
long-styled form stand clear above the anthers, and the whole
pistil is longer than that of the short-styled, in about the ratio
of 3 to 2.

The stigmas of the short-styled form are seated beneath the
anthers, and they are considerably shorter than those of the
long-styled form. This latter difference is the most important
one of any between the two forms.

Gilia (Ipomopsis) pulchella vel aggregata (Polemoni-aceæ).

Professor Asa Gray remarks with respect to this plant : " the
tendency to dimorphism, of which there are traces, or perhaps
rather incipient manifestations in various portions of the genus,
is most marked in *G. aggregata.*"[*] He sent me some dried
flowers, and I procured others from Kew. They differ greatly
in size, some being nearly twice as long as others (viz. as 30 to
17), so that it was not possible to compare, except by calculation,
the absolute length of the organs from different plants. Moreover, the relative position of the stigmas and anthers is variable :
in some long-styled flowers the stigmas and anthers were exserted only just beyond the throat of the corolla; whilst in
others they were exserted as much as $\frac{4}{10}$ of an inch. I suspect
also that the pistil goes on growing for some time after the
anthers have dehisced. Nevertheless it is possible to class the
flowers under two forms. In some of the long-styled, the length
of pistil to that of the short-styled was as 100 to 82; but this
result was gained by reducing the size of the corollas to the
same scale. In another pair of flowers the difference in length
between the pistils of the two forms was certainly greater, but
they were not actually measured. In the short-styled flowers
whether large or small, the stigma is seated low down within
the tube of the corolla. The papillæ on the long-styled stigma
are longer than those on the short-styled, in the ratio of 100 to
40. The filaments in some of the short-styled flowers were, to
those of the long-styled, as 100 to 25 in length, the free, or

---

[*] 'Proc. American Acad. of Arts and Sciences,' June 14, 1870, p. 275.

unattached portion being alone measured; but this ratio cannot be trusted, owing to the great variability of the stamens. The mean diameter of eleven pollen-grains from long-styled flowers, and of twelve from the short-styled, was exactly the same. It follows from these several statements, that the difference in length and state of surface of the stigmas in the flowers is the sole reliable evidence that this species is heterostyled; for it would be rash to trust to the difference in the length of the pistils, seeing how variable they are. I should have left the case altogether doubtful, had it not been for the observations on the following species; and these leave little doubt on my mind that the present plant is truly heterostyled. Professor Gray informs me that in another species, *G. coronopifolia*, belonging to the same section of the genus, he can see no sign of dimorphism.

GILIA (LEPTOSIPHON) MICRANTHA.

A few flowers sent me from Kew had been somewhat injured, so that I cannot say anything positively with respect to the position and relative length of the organs in the two forms. But their stigmas differed almost exactly in the same manner as in the last species; the papillæ on the long-styled stigma being longer than those on the short-styled, in the ratio of 100 to 42. My son measured nine pollen-grains from the long-styled, and the same number from the short-styled form; and the mean diameter of the former was to that of the latter as 100 to 81. Considering this difference, as well as that between the stigmas of the two forms, there can be no doubt that this species is heterostyled. So probably is *Gilia nudicaulis*, which likewise belongs to the Leptosiphon section of the genus, for I hear from Professor Asa Gray that in some individuals the style is very long, with the stigma more or less exserted, whilst in others it is deeply included within the tube; the anthers being always seated in the throat of the corolla.

PHLOX SUBULATA (POLEMONIACEÆ).

Professor Asa Gray informs me that the greater number of the species in this genus have a long pistil, with the stigma more or less exserted; whilst several other species, especially the annuals, have a short pistil seated low down within the tube of the corolla. In all the species the anthers are arranged one

below the other, the uppermost just protruding from the throat
of the corolla. In *Phlox subulata* alone he has " seen both long
and short styles; and here the short-styled plant has (irrespec-
tive of this character) been described as a distinct species (*P.
nivalis, P. Hentzii*), and is apt to have a pair of ovules in each
cell, while the long-styled *P. subulata* rarely shows more than
one."* Some dried flowers of both forms were sent me by him,
and I received others from Kew, but I have failed to make
out whether the species is heterostyled. In two flowers of
nearly equal size, the pistil of the long-styled form was twice as
long as that of the short-styled; but in other cases the difference was not nearly so great. The stigma of the long-styled
pistil stands nearly in the throat of the corolla; whilst in the
short-styled it is placed low down—sometimes very low down
in the tube, for it varies greatly in position. The stigma is
more papillose, and of greater length (in one instance in the
ratio of 100 to 67), in the short-styled flowers than in the
long-styled. My son measured twenty pollen-grains from a
short-styled flower, and nine from a long-styled, and the
former were in diameter to the latter as 100 to 93; and this
difference accords with the belief that the plant is heterostyled. But the grains from the short-styled varied much in
diameter. He afterwards measured ten grains from a distinct
long-styled flower, and ten from another plant of the same form,
and these grains differed in diameter in the ratio of 100 to 90.
The mean diameter of these two lots of twenty grains was to
that of twelve grains from another short-styled flower as 100 to
75: here, then, the grains from the short-styled form were considerably smaller than those from the long-styled, which is the
reverse of what occurred in the former instance, and of what is
the general rule with heterostyled plants. The whole case is
perplexing in the highest degree, and will not be understood
until experiments are tried on living plants. The greater length,
and more papillose condition of the stigma in the short-styled
than in the long-styled flowers, looks as if the plant was heterostyled; for we know that with some species—for instance, Leucosmia and certain Rubiaceæ—the stigma is longer and more
papillose in the short-styled form, though the reverse of this
holds good in Gilia, a member of the same family with Phlox.
The similar position of the anthers in the two forms is some-

---

* ' Proc. American Acad. of Arts and Sciences,' June 14, 1870, p. 248.

what opposed to the present species being heterostyled; as is
the great difference in the length of the pistil in several short-
styled flowers. But the extraordinary variability in diameter of
the pollen-grains, and the fact that in one set of flowers the
grains from the long-styled flowers were larger than those from
the short-styled, is strongly opposed to the belief that *Phlox
subulata* is heterostyled. Possibly this species was once heterostyled, but is now becoming sub-diœcious; the short-styled
plants having been rendered more feminine in nature. This
would account for their ovaries usually containing more ovules,
and for the variable condition of their pollen-grains. Whether
the long-styled plants are now changing their nature, as would
appear to be the case from the variability of their pollen-grains,
and are becoming more masculine, I will not pretend to conjecture; they might remain as hermaphrodites, for the co-
existence of hermaphrodite and female plants of the same
species is by no means a rare event.

ERYTHROXYLUM [SP. ?] (ERYTHROXYLIDÆ).

Fritz Müller sent me from South Brazil dried flowers of this
tree, together with the accompanying drawings, which show the
two forms, magnified about five times, with the petals removed.

Fig. 8.

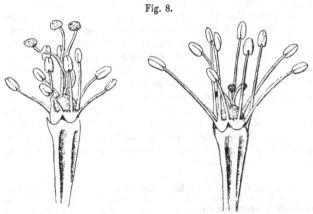

Long-styled form.   Short-styled form.
From a sketch by Fritz Müller, magnified five times.
ERYTHROXYLON [sp. ?].

In the long-styled form the stigmas project above the anthers, and the styles are nearly twice as long as those of the short-styled form, in which the stigmas stand beneath the anthers. The stigmas in many, but not in all the short-styled flowers are larger than those in the long-styled. The anthers of the short-styled flowers stand on a level with the stigmas of the other form; but the stamens are longer by only one-fourth or one-fifth of their own length than those of the long-styled. Consequently the anthers of the latter do not stand on a level with, but rather above the stigmas of the other form. Differently from what occurs in the following closely allied genus, Sethia, the stamens are of nearly equal length in the flowers of the same form. The pollen-grains of the short-styled flowers, measured in their dry state, are a little larger than those from the long-styled flowers in about the ratio of 100 to 93.*

SETHIA ACUMINATA (ERYTHROXYLIDÆ).

Mr. Thwaites pointed out several years ago† that this plant exists under two forms, which he designated as *forma stylosa et staminea;* and the flowers sent to me by him are clearly heterostyled. In the long-styled form the pistil is nearly twice as long, and the stamens half as long as the corresponding organs in the short-styled form. The stigmas of the long-styled seem rather smaller than those of the short-styled. All the stamens in the short-styled flowers are of nearly equal length, whereas in long-styled they differ in length being alternately a little longer and shorter; and this difference in the stamens of the two forms is probably related, as we shall hereafter see in the case of the short-styled flowers of *Lythrum salicaria*, to the manner in which insects can best transport pollen from the long-styled flowers to the stigmas of the short-styled. The pollen-grains from the short-styled flowers, though variable in size, are to those of the long-styled, as far as I could make out, as 100 to 83 in their longer diameter. *Sethia obtusifolia* is heterostyled like *S. acuminata*.

---

* F. Müller remarks in his letter to me that the flowers, of which he carefully examined many specimens, are curiously variable in the number of their parts: 5 sepals and petals, 10 stamens and 3 pistils are the prevailing numbers; but the sepals and petals often vary from 5 to 7; the stamens from 10 to 14, and the pistils from 3 to 4.

† 'Enumeratio Plantarum Zeylaniæ,' 1864, p. 54.

CRATOXYLON FORMOSUM (HYPERICINEÆ).

Mr. Thiselton Dyer remarks that this tree, an inhabitant of Malacca and Borneo, appears to be heterostyled.* He sent me dried flowers, and the difference between the two forms is conspicuous. In the long-styled form the pistils are in length to those of the short-styled as 100 to 40, with their globular stigmas about twice as thick. These stand just above the numerous anthers and a little beneath the tips of the petals. In the short-styled form the anthers project high above the pistils, the stigmas of which diverge between the three bundles of stamens, and stand only a little above the tips of the sepals. The stamens in this form are to those of the long-styled as 100 to 86 in length; and therefore they do not differ so much in length as do the pistils. Ten pollen-grains from each form were measured, and those from the short-styled were to those from the long-styled as 100 to 86 in diameter. This plant, therefore, is in all respects a well-characterised heterostyled species.

ÆGIPHILA ELATA (VERBENACEÆ).

Mr. Bentham was so kind as to send me dried flowers of this species and of *Æ. mollis*, both inhabitants of South America. The two forms differ conspicuously, as the deeply bifid stigma of the one, and the anthers of the other project far above the mouth of the corolla. In the long-styled form of the present species, the style is twice and a half as long as that of the short-styled. The divergent stigmas of the two forms do not differ much in length, nor as far as I could perceive in their papillæ. In the long-styled flowers the filaments adhere to the corolla close up to the anthers, which are enclosed some way down within the tube. In the short-styled flowers the filaments are free above the point where the anthers are seated in the other form, and they project from the corolla to an equal height with that of the stigmas in the long-styled flowers. It is often difficult to measure with accuracy pollen-grains, which have long been dried and then soaked in water; but they here manifestly differed greatly in size. Those from the short-styled flowers were to those from the long-styled in diameter in

---

* 'Journal of Botany,' London, 1872, p. 26.

about the ratio of 100 to 62. The two forms of *Æ. mollis* present a like difference in the length of their pistils and stamens.

ÆGIPHILA OBDURATA.

Flowers of this bush were sent me from St. Catharina in Brazil, by Fritz Müller, and were named for me at Kew. They appeared at first sight grandly heterostyled, as the stigma of the long-styled form projects far out of the corolla, whilst the anthers are seated halfway down within the tube; whereas in the short-styled form the anthers project from the corolla and the stigma is enclosed in the tube at nearly the same level with the anthers of the other form. The pistil of the long-styled is to that of the short-styled as 100 to 60 in length, and the stigmas, taken by themselves, as 100 to 55. Nevertheless, this plant cannot be heterostyled. The anthers in the long-styled form are brown, tough, and fleshy, and less than half the length of those in the short-styled form, strictly as 44 to 100 ; and what is much more important, they were in a rudimentary condition in the two flowers examined by me, and did not contain a single grain of pollen. In the short-styled form, the divided stigma, which as we have seen is much shortened, is thicker and more fleshy than the stigma of the long-styled, and is covered with small irregular projections, formed of rather large cells. It had the appearance of having suffered from hypertrophy, and is probably incapable of fertilisation. If this be so the plant is diœcious, and judging from the two species previously described, it probably was once heterostyled, and has since been rendered diœcious by the pistil in the one form, and the stamens in the other having become functionless and reduced in size. It is, however, possible that the flowers may be in the same state as those of the common thyme and of several other Labiatæ, in which females and hermaphrodites regularly co-exist. Fritz Müller, who thought that the present plant was heterostyled, as I did at first, informs me that he found bushes in several places growing quite isolated, and that these were completely sterile; whilst two plants growing close together were covered with fruit. This fact agrees better with the belief that the species is diœcious than that it consists of hermaphrodites and females; for if any one of the isolated plants had been an hermaphrodite, it would probably have produced some fruit.

Rubiaceæ.

This great natural family contains a much larger number of heterostyled genera than any other one, as yet known.

Mitchella repens.—Prof. Asa Gray sent me several living plants collected when out of flower, and nearly half of these proved long-styled, and the other half short-styled. The white flowers, which are fragrant and which secrete plenty of nectar, always grow in pairs with their ovaries united, so that the two together produce " a berry-like double drupe."* In my first series of experiments (1864) I did not suppose that this curious arrangement of the flowers would have any influence on their fertility; and in several instances only one of the two flowers in a pair was fertilised; and a large proportion or all of these failed to produce berries. In the ensuing year both flowers of each pair were invariably fertilised in the same manner; and the latter experiments alone serve to show the proportion of flowers which yield berries, when legitimately and illegitimately fertilised; but for calculating the average number of seeds per berry I have used those produced during both seasons.

In the long-styled flowers the stigma projects just above the bearded throat of the corolla, and the anthers are seated some way down the tube. In the short-styled flowers these organs occupy reversed positions. In this latter form the fresh pollen-grains are a little larger and more opaque than those of the long-styled form. The results of my experiments are given in the following table.

---

* A. Gray, 'Manual of the Bot. of the N. United States,' 1856, p. 172.

TABLE 21.

Mitchella repens.

Nature of Union.	Number of Pairs of Flowers fertilised during the second Season.	Number of Drupes produced during the second Season.	Average Number of good Seeds per Drupe in all the Drupes during the two Seasons.
Long-styled flowers, by pollen of short-styled. Legitimate union	9	8	4·6
Long-styled flowers, by own-form pollen. Illegitimate union . .	8	3	2·2
Short-styled flowers, by pollen of long-styled. Legitimate union.	8	7	4·1
Short-styled flowers, by own-form pollen. Illegitimate union . .	9	0	2·0
The two legitimate unions together	17	15	4·4
The two illegitimate unions together	17	3	2·1

It follows from this table that 88 per cent. of the paired flowers of both forms, when legitimately fertilised, yielded double berries, nineteen of which contained on an average 4·4 seeds, with a maximum in one of 8 seeds. Of the illegitimately fertilised paired flowers only 18 per cent. yielded berries, six of which contained on an average only 2·1 seeds, with a maximum in one of 4 seeds. Thus the two legitimate unions are more fertile than the two illegitimate, according to the proportion of flowers which yielded berries, in the ratio of 100 to 20; and according to the average number of contained seeds as 100 to 47.

Three long-styled and three short-styled plants were protected under separate nets, and they produced altogether only 8 berries, containing on an average only

1·5 seed.   Some additional berries were produced
which contained no seeds.   The plants thus treated were
therefore excessively sterile, and their slight degree of
fertility may be attributed in part to the action of the
many individuals of Thrips which haunted the flowers.
Mr. J. Scott informs me that a single plant (probably
a long-styled one), growing in the Botanic Gardens at
Edinburgh, which no doubt was freely visited by insects, produced plenty of berries, but how many of
them contained seeds was not observed.

BORRERIA, NOV. SP. NEAR VALERIANOIDES (RUBIACEÆ).

Fritz Müller sent me seeds of this plant, which is
extremely abundant in St. Catharina, in South Brazil ;
and ten plants were raised, consisting of five long-
styled and five short-styled.   The pistil of the long-
styled flowers projects just beyond the mouth of the
corolla, and is thrice as long as that of the short-
styled, and the divergent stigmas are likewise rather
larger.   The anthers in the long-styled form stand
low down within the corolla, and are quite hidden.
In the short-styled flowers the anthers project just
above the mouth of the corolla, and the stigma stands
low down within the tube.   Considering the great
difference in the length of the pistils in the two forms,
it is remarkable that the pollen-grains differ very little
in size, and Fritz Müller was struck with the same
fact.   In a dry state the grains from the short-styled
flowers could just be perceived to be larger than those
from the long-styled, and when both were swollen by
immersion in water, the former were to the latter in
diameter in the ratio of 100 to 92.   In the long-styled
flowers beaded hairs almost fill up the mouth of the
corolla and project above it ; they therefore stand
above the anthers and beneath the stigma.   In the

short-styled flowers a similar brush of hairs is situated low down within the tubular corolla, above the stigma and beneath the anthers. The presence of these beaded hairs in both forms, though occupying such different positions, shows that they are probably of considerable functional importance. They would serve to guard the stigma of each form from its own pollen; but in accordance with Prof. Kerner's view* their chief use probably is to prevent the copious nectar being stolen by small crawling insects, which could not render any service to the species by carrying pollen from one form to the other.

The flowers are so small and so crowded together that I was not willing to expend time in fertilising them separately; but I dragged repeatedly heads of short-styled flowers over three long-styled flower-heads, which were thus legitimately fertilised; and they produced many dozen fruits, each containing two good seeds. I fertilised in the same manner three heads on the same long-styled plant with pollen from another long-styled plant, so that these were fertilised illegitimately, and they did not yield a single seed. Nor did this plant, which was of course protected by a net, bear spontaneously any seeds. Nevertheless another long-styled plant, which was carefully protected, produced spontaneously a very few seeds; so that the long-styled form is not always quite sterile with its own pollen.

FARAMEA [SP. ?] (RUBIACEÆ).

Fritz Müller has fully described the two forms of this remarkable plant, an inhabitant of South Brazil.† In

---

* 'Die Schutzmittel der Blüthen gegen unberufene Gäste,'
1876, p. 37.

† 'Bot. Zeitung,' Sept. 10, 1869,
p. 606.

the long-styled form the pistil projects above the corolla, and is almost exactly twice as long as that of the short-styled, which is included within the tube. The former is divided into two rather short and broad stigmas, whilst the short-styled pistil is divided into two long, thin, sometimes much curled stigmas. The stamens of each form correspond in height or length with the pistils of the other form. The anthers of the short-styled form are a little larger than those of the long-styled; and their pollen-grains are to those of the other form as 100 to 67 in diameter. But the pollen-grains of the two forms differ in a much more remarkable manner, of which no other

Fig. 9.

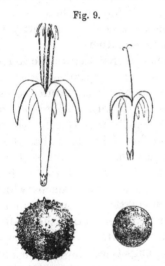

'Short-styled form.   Long-styled form.
Outlines of flowers from dried specimens.   Pollen-grains, magnified 180
times, by Fritz Müller.

FARAMEA [sp. ?].

instance is known; those from the short-styled flowers being covered with sharp points; the smaller ones

K

from the long-styled being quite smooth. Fritz Müller remarks that this difference between the pollen-grains of the two forms is evidently of service to the plant; for the grains from the projecting stamens of the short-styled form, if smooth, would have been liable to be blown away by the wind, and would thus have been lost; but the little points on their surfaces cause them to cohere, and at the same time favour their adhesion to the hairy bodies of insects, which merely brush against the anthers of these stamens whilst visiting the flowers. On the other hand, the smooth grains of the long-styled flowers are safely included within the tube of the corolla, so that they cannot be blown away, but are almost sure to adhere to the proboscis of an entering insect, which is necessarily pressed close against the enclosed anthers.

It may be remembered that in the long-styled form of *Linum perenne* each separate stigma rotates on its own axis, when the flower is mature, so as to turn its papillose surface outwards. There can be no doubt that this movement, which is confined to the long-styled form, is effected in order that the proper surface of the stigma should receive pollen brought by insects from the other form. Now with Faramea, as Fritz Müller shows, it is the stamens which rotate on their axes in one of the two forms, namely, the short-styled, in order that their pollen should be brushed off by insects and transported to the stigmas of the other form. In the long-styled flowers the anthers of the short enclosed stamens do not rotate on their axes, but dehisce on their inner sides, as is the common rule with the Rubiaceæ; and this is the best position for the adherence of the pollen-grains to the proboscis of an entering insect. Fritz Müller therefore infers that as the plant became heterostyled, and as the stamens of the short-styled form increased in length, they gradually acquired the highly beneficial power of rotating on their own axes. But he has further shown, by the careful examination of many flowers, that this power has not as yet been perfected; and, consequently, that a certain proportion of the pollen is rendered useless, namely, that from the anthers which do not rotate properly. It thus appears that the development of the plant has not as yet been completed; the stamens have indeed acquired their proper length, but not their full and perfect power of rotation.*

The several points of difference in structure between the two forms of Faramea are highly remarkable. Until within a recent period, if any one had been shown two plants which differed in a uniform manner in the length of their stamens and pistils,—in the form of their stigmas,—in the manner of dehiscence and slightly in the size of their anthers,—and to an extraordinary degree in the diameter and structure of their pollen-grains, he would have declared it impossible that the two could have belonged to one and the same species.

SUTERIA (species unnamed in the herbarium at Kew)
(RUBIACEÆ).

I owe to the kindness of Fritz Müller dried flowers of this plant from St. Catharina, in Brazil. In the long-styled form the stigma stands in the mouth of the corolla, above the anthers,

---

* Fritz Müller gives another instance of the want of absolute perfection in the flowers of another member of the Rubiaceæ, namely, *Posoqueria fragrans*, which is adapted in a most wonderful manner for cross-fertilisation by the agency of moths. (See ‘Bot. Zeitung,’ 1866, No. 17.) In accordance with the nocturnal habits of these insects, most of the flowers open only during the night; but some open in the day, and the pollen of such flowers is robbed, as Fritz Müller has often seen, by humble-bees and other insects. without any benefit being thus conferred on the plant.

which latter are enclosed within the tube, but only a short way down. In the short-styled form the anthers are placed in the mouth of the corolla above the stigma, which occupies the same position as the anthers in the other form, being seated only a short way down the tube. Therefore the pistil of the long-styled form does not exceed in length that of the short-styled in nearly so great a degree as in many other Rubiaceæ. Nevertheless there is a considerable difference in the size of the pollen-grains in the two forms; for, as Fritz Müller informs me, those of the short-styled are to those of the long-styled as 100 to 75 in diameter.

HOUSTONIA CŒRULEA (RUBIACEÆ).

Prof. Asa Gray has been so kind as to send me an abstract of some observations made by Dr. Rothrock on this plant. The pistil is exserted in the one form and the stamens in the other, as has long been observed. The stigmas of the long-styled form are shorter, stouter, and far more hispid than in the other form. The stigmatic hairs or papillæ on the former are ·04 mm., and on the latter only ·023 mm. in length. In the short-styled form the anthers are larger, and the pollen-grains, when distended with water, are to those from the long-styled form as 100 to 72 in diameter.

Selected capsules from some long-styled plants growing in the Botanic Gardens at Cambridge, U.S., near where plants of the other form grew, contained on an average 13 seeds; but these plants must have been subjected to unfavourable conditions, for some long-styled plants in a state of nature yielded an average of 21·5 seeds per capsule. Some short-styled plants, which had been planted by themselves in the Botanic Gardens, where it was not likely that they would have been visited by insects that had previously visited long-styled plants, produced capsules, eleven of which were wholly sterile, but one contained 4, and another 8 seeds. So that the short-styled form seems to be very sterile with its own pollen. Prof. Asa Gray informs me that the other North American species of this genus are likewise heterostyled.

OLDENLANDIA [SP. ?] (RUBIACEÆ).

Mr. J. Scott sent me from India dried flowers of a hetero-styled species of this genus, which is closely allied to the last.

The pistil in the long-styled flowers is longer by about a quarter of its length, and the stamens shorter in about the same proportion, than the corresponding organs in the short-styled flowers. In the latter the anthers are longer, and the divergent stigmas decidedly longer and apparently thinner than in the long-styled form. Owing to the state of the specimens, I could not decide whether the stigmatic papillæ were longer in the one form than in the other. The pollen-grains, distended with water, from the short-styled flowers were to those from the long-styled as 100 to 78 in diameter, as deduced from the mean of ten measurements of each kind.

HEDYOTIS [SP. ?] (RUBIACEÆ).

Fritz Müller sent me from St. Catharina, in Brazil, dried flowers of a small delicate species, which grows on wet sand near the edges of fresh-water pools. In the long-styled form the stigma projects above the corolla, and stands on a level with the projecting anthers of the short-styled form; but in the latter the stigmas stand rather beneath the level of the anthers in the other or long-styled form, these being enclosed within the tube of the corolla. The pistil of the long-styled form is nearly thrice as long as that of the short-styled, or, speaking strictly, as 100 to 39; and the papillæ on the stigma of the former are broader, in the ratio of 4 to 3, but whether longer than those of the short-styled, I could not decide. In the short-styled form, the anthers are rather larger, and the pollen-grains are to those from the long-styled flowers, as 100 to 88 in diameter. Fritz Müller sent me a second, small-sized species, which is likewise heterostyled.

COCCOCYPSELUM [SP. ?] (RUBIACEÆ).

Fritz Müller also sent me dried flowers of this plant from St. Catharina, in Brazil. The exserted stigma of the long-styled form stands a little above the level of the exserted anthers of the short-styled form; and the enclosed stigma of the latter also stands a little above the level of the enclosed anthers in the long-styled form. The pistil of the long-styled is about twice as long as that of the short-styled, with its two stigmas considerably longer, more divergent, and more curled. Fritz Müller informs me that he could detect no difference in the size of the pollen-grains in the two forms.　Nevertheless, there can be no doubt that this plant is heterostyled.

Lipostoma [sp. ?] (Rubiaceæ).

Dried flowers of this plant, which grows in small wet ditches in St. Catharina, in Brazil, were likewise sent me by Fritz Müller.　In the long-styled form the exserted stigma stands rather above the level of the exserted anthers of the other form; whilst in the short-styled form it stands on a level with the anthers of the other form.　So that the want of strict corre-spondence in height between the stigmas and anthers in the two forms is reversed, compared with what occurs in Hedyotis.　The long-styled pistil is to that of the short-styled as 100 to 36 in length; and its divergent stigmas are longer by fully one-third of their own length than those of the short-styled form.　In the latter the anthers are a little larger, and the pollen-grains are as 100 to 80 in diameter, compared with those from the long-styled form.

Cinchona micrantha (Rubiaceæ).

Dried specimens of both forms of this plant were sent me from Kew.*　In the long-styled form the apex of the stigma stands just beneath the bases of the hairy lobes of the corolla; whilst the summits of the anthers are seated about halfway down the tube.　The pistil is in length as 100 to 38 to that of the short-styled form.　In the latter the anthers occupy the same position as the stigma of the other form, and they are con-siderably longer than those of the long-styled form.　As the summit of the stigma in the short-styled form stands beneath the bases of the anthers, which are seated halfway down the corolla, the style has been extremely shortened in this form; its length to that of the long-styled being, in the specimens examined, only as 5·3 to 100!　The stigma, also, in the short-styled form is very much shorter than that in the long-styled, in the ratio of 57 to 100.　The pollen-grains from the short-

---

* My attention was called to this plant by a drawing copied from Howard's ' Quinologia,' Tab. 3, given by Mr. Markham in his ' Travels in Peru,' p. 539.

styled flowers, after having been soaked in water, were rather
larger—in about the ratio of 100 to 91—than those from the long-
styled flowers, and they were more triangular, with the angles
more prominent. As all the grains from the short-styled flowers
were thus characterised, and as they had been left in water for
three days, I am convinced that this difference in shape in the
two sets of grains cannot be accounted for by unequal distension
with water.

Besides the several Rubiaceous genera already mentioned,
Fritz Müller informs me that two or three species of Psychotria
and *Rudgea eriantha*, natives of St. Catharina, in Brazil, are
heterostyled, as is *Manettia bicolor*. I may add that I formerly
fertilised with their own pollen several flowers on a plant of
this latter species in my hothouse, but they did not set a single
fruit. From Wight and Arnott's description, there seems to be
little doubt that Knoxia in India is heterostyled; and Asa Gray
is convinced that this is the case with Diodia and Spermacoce
in the United States. Lastly, from Mr. W. W. Bailey's description,* it appears that the Mexican *Bouvardia leiantha is* heterostyled.

Altogether we now know of 17 heterostyled genera
in the great family of the Rubiaceæ; though more
information is necessary with respect to some of them,
more especially those mentioned in the last para-
graph, before we can feel absolutely safe. In the
'Genera Plantarum,' by Bentham and Hooker, the
Rubiaceæ are divided into 25 tribes, containing 337
genera; and it deserves notice that the genera now
known to be heterostyled are not grouped in one or
two of these tribes, but are distributed in no less than
eight of them. From this fact we may infer that
most of the genera have acquired their heterostyled
structure independently of one another; that is, they
have not inherited this structure from some one or
even two or three progenitors in common. It further

---

* 'Bull. of the Torrey Bot. Club,' 1876, p. 106.

deserves notice that in the homostyled genera, as I
am informed by Professor Asa Gray, the stamens are
either exserted or are included within the tube of the
corolla, in a nearly constant manner; so that this
character, which is not even of specific value in the
heterostyled species, is often of generic value in other
members of the family.

CHAPTER IV.

Heterostyled Trimorphic Plants.

Lythrum salicaria—Description of the three forms—Their power and complex manner of fertilising one another—Eighteen different unions possible—Mid-styled form eminently feminine in nature—Lythrum Græfferi likewise trimorphic—L. thymifolia dimorphic—L. hyssopifolia homostyled—Nesæa verticillata trimorphic—Lagerstrœmia, nature doubtful—Oxalis, trimorphic species of—O. Valdiviana—O. Regnelli, the illegitimate unions quite barren—O. speciosa—O. sensitiva—Homostyled species of Oxalis—Pontederia, the one monocotyledonous genus known to include heterostyled species.

In the previous chapters various heterostyled dimorphic plants have been described, and now we come to heterostyled trimorphic plants, or those which present three forms. These have been observed in three families, and consist of species of Lythrum and of the allied genus Nesæa, of Oxalis and Pontederia. In their manner of fertilisation these plants offer a more remarkable case than can be found in any other plant or animal.

Lythrum salicaria.—The pistil in each form differs from that in either of the other forms, and in each there are two sets of stamens different in appearance and function. But one set of stamens in each form corresponds with a set in one of the other two forms. Altogether this one species includes three females or female organs and three sets of male organs, all as distinct from one another as if they belonged to different species; and if smaller functional differences are considered, there are five distinct sets of males. Two of the three hermaphrodites must coexist, and pollen must be carried by insects reciprocally from one to the other, in order that either of the two should be fully fertile; but unless all three forms coexist, two sets of stamens will be wasted, and the organisation of the species, as a whole, will be incomplete. On the other hand, when all three hermaphrodites coexist, and pollen is carried from one to the other, the scheme is perfect; there is no waste of pollen and no false co-adaptation. In short, nature has ordained a most complex marriage-arrangement, namely a triple union between three hermaphrodites,—each hermaphrodite being in its female organ quite distinct from the other two hermaphrodites and partially distinct in its male organs, and each furnished with two sets of males.

The three forms may be conveniently called, from the unequal lengths of their pistils, the *long-styled*, *mid-styled*, and *short-styled*. The stamens also are of unequal lengths, and these may be called the *longest*, *mid-length*, and *shortest*. Two sets of stamens of different length are found in each form. The existence of the three forms was first observed by Vaucher,* and subsequently more carefully by Wirtgen; but these botanists, not being guided by any theory or even suspicion of their functional differences, did not perceive some of the most curious points of difference in their structure. I will first briefly describe the three forms by the aid of the accompanying diagram, which shows the flowers, six times magnified, in their natural position, with their petals and calyx on the near side removed.

---

* 'Hist. Phys. des Plantes d'Europe,' tom. ii. 1841, p. 371. Wirtgen,"Ueber *Lythrum salicaria* und dessen Formen," 'Verhand. des naturhist. Vereins für preuss. Rheinl.' 5. Jahrgang, 1848, S. 7.

Fig. 10.

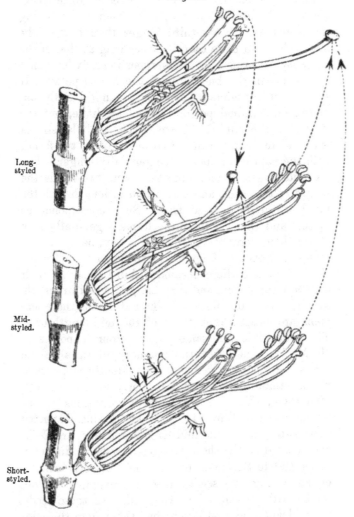

Long-
styled

Mid-
styled.

Short-
styled.

Diagram of the flowers of the three forms of *Lythrum salicaria*, in their natural
position, with the petals and calyx removed on the near side: enlarged six times.
The dotted lines with the arrows show the directions in which pollen must be
carried to each stigma to ensure full fertility.

Long-styled form.—This form can be at once recognised by the length of the pistil, which is (including the ovarium) fully one-third longer than that of the mid-styled, and more than thrice as long as that of the short-styled form. It is so disproportionately long, that it projects in the bud through the folded petals. It stands out considerably beyond the mid-length stamens; its terminal portion depends a little, but the stigma itself is slightly upturned. The globular stigma is considerably larger than that of the other two forms, with the papillæ on its surface generally longer. The six mid-length stamens project about two-thirds the length of the pistil, and correspond in length with the pistil of the mid-styled form. Such correspondence in this and the two following forms is generally very close; the difference, where there is any, being usually in a slight excess of length in the stamens. The six shortest stamens lie concealed within the calyx; their ends are turned up, and they are graduated in length, so as to form a double row. The anthers of these stamens are smaller than those of the mid-length ones. The pollen is of the same yellow colour in both sets. H. Müller* measured the pollen-grain in all three forms, and his measurements are evidently more trustworthy than those which I formerly made, so I will give them. The numbers refer to divisions of the micrometer equalling $\frac{1}{300}$ ..... The grains, distended with water, from the mid-length stamens are 7–7$\frac{1}{2}$, and those from the shortest stamens 6–6$\frac{1}{2}$ in diameter, or as 100 to 86. The capsules of this form contain on an average 93 seeds: how this average was obtained will presently be explained. As these seeds, when cleaned, seemed larger than those from the mid-

---

* 'Die Befruchtung der Blumen,' 1873, p. 193.

styled or short-styled forms, 100 of them were placed in a good balance, and by the double method of weighing were found to equal 121 seeds of the mid-styled or 142 of the short-styled; so that five long-styled seeds very nearly equal six mid-styled or seven short-styled seeds.

Mid-styled form.—The pistil occupies the position represented in the diagram, with its extremity considerably upturned, but to a variable degree; the stigma is seated between the anthers of the longest and the shortest stamens. The six longest stamens correspond in length with the pistil of the long-styled form; their filaments are coloured bright pink; the anthers are dark-coloured, but from containing bright-green pollen and from their early dehiscence they appear emerald-green. Hence in general appearance these stamens are remarkably dissimilar from the mid-length stamens of the long-styled form. The six shortest stamens are enclosed within the calyx, and resemble in all respects the shortest stamens of the long-styled form; both these sets correspond in length with the short pistil of the short-styled form. The green pollen-grains of the longest stamens are 9–10 in diameter, whilst the yellow grains from the shortest stamens are only 6; or as 100 to 63. But the pollen-grains from different plants appeared to me, in this case and others, to be in some degree variable in size. The capsules contain on an average 130 seeds; but perhaps, as we shall see, this is rather too high an average. The seeds themselves, as before remarked, are smaller than those of the long-styled form.

Short-styled form.—The pistil is here very short, not one-third of the length of that of the long-styled form. It is enclosed within the calyx, which, differently from that in the other two forms, does not enclose any an-

thers. The end of the pistil is generally bent upwards
at right angles. The six longest stamens, with their
pink filaments and green pollen, resemble the corresponding stamens of the mid-styled form. But according to H. Müller, their pollen-grains are a little larger,
viz. $9\frac{1}{2}$–$10\frac{1}{2}$, instead of 9–10 in diameter. The six
mid-length stamens, with their uncoloured filaments
and yellow pollen, resemble in the size of their pollen-
grains and in all other respects the corresponding
stamens of the long-styled form. The difference in
diameter between the grains from the two sets of
anthers in the short-styled form is as 100 to 73.
The capsules contain fewer seeds on an average than
those of either of the preceding forms, namely 83·5;
and the seeds are considerably smaller. In this latter
respect, but not in number, there is a gradation
parallel to that in the length of the pistil, the long-
styled having the largest seeds, the mid-styled the
next in size, and the short-styled the smallest.

We thus see that this plant exists under three
female forms, which differ in the length and curvature of the style, in the size and state of the stigma,
and in the number and size of the seed. There are
altogether thirty-six males or stamens, and these can
be divided into three sets of a dozen each, differing
from one another in length, curvature, and colour of
the filaments—in the size of the anthers, and especially
in the colour and diameter of the pollen-grains. Each
form bears half-a-dozen of one kind of stamens and
half-a-dozen of another kind, but not all three kinds.
The three kinds of stamens correspond in length with
the three pistils : the correspondence is always between
half of the stamens in two of the forms with the pistil
of the third form. The following table of the diameters
of the pollen-grains, after immersion in water, from

both sets of stamens in all three forms is copied from
H. Müller; they are arranged in the order of their
size :—

Pollen-grains from longest stamens of short-styled form			$9\frac{1}{2}$ to $10\frac{1}{2}$
,,	,, ,, ,, mid-styled	,,	9 ,, 10 ,
,,	,, mid-length stamens of long-styled	,,	7 ,, $7\frac{1}{2}$
,,	,, ,, ,, short-styled	,,	7 ,, $7\frac{1}{2}$
,,	,, shortest stamens of long-styled	,,	6 ,, $6\frac{1}{4}$
,,	,, ,, ,, mid-styled	,,	6 ,, 6

We here see that the largest pollen-grains come from
the longest stamens, and the least from the shortest;
the extreme difference in diameter between them
being as 100 to 60.

The average number of seeds in the three forms was
ascertained by counting them in eight fine selected
capsules taken from plants growing wild, and the
result was, as we have seen, for the long-styled (neg-
lecting decimals) 93, mid-styled 130, and short-styled
83. I should not have trusted in these ratios had I
not possessed a number of plants in my garden which,
owing to their youth, did not yield the full complement of seed, but were of the same age and grew
under the same conditions, and were freely visited by
bees. I took six fine capsules from each, and found
the average to be for the long-styled 80, for the mid-
styled 97, and for the short-styled 61. Lastly, legitimate unions effected by me between the three forms
gave, as may be seen in the following tables, for the
long-styled an average of 90 seeds, for the mid-styled
117, and for the short-styled 71. So that we have
good concurrent evidence of a difference in the average
production of seed by the three forms. To show that
the unions effected by me often produced their full
effect and may be trusted, I may state that one mid-
styled capsule yielded 151 good seeds, which is the
same number as in the finest wild capsule which I

examined.   Some artificially fertilised short- and long-
styled capsules produced a greater number of seeds than
was ever observed by me in wild plants of the same
forms, but then I did not examine many of the latter.
This plant, I may add, offers a remarkable instance, how
profoundly ignorant we are of the life-conditions of a
species.   Naturally it grows " in wet ditches, watery
places, and especially on the banks of streams," and
though it produces so many minute seeds, it never
spreads on the adjoining land ; yet, when planted in my
garden, on clayey soil lying over chalk, and which is so
dry that a rush cannot be found, it thrives luxuriantly,
grows to above 6 feet in height, produces self-sown
seedlings, and (which is a severer test) is as fertile as
in a state of nature.   Nevertheless it would be almost
a miracle to find this plant growing spontaneously on
such land as that in my garden.

According to Vaucher and Wirtgen, the three forms
coexist in all parts of Europe.   Some friends gathered
for me in North Wales a number of twigs from
separate plants growing near one another, and classified them.   My son did the same in Hampshire, and
here is the result :—

TABLE 22.

	Long-styled.	Mid-styled.	Short-styled.	Total.
North Wales . . .	95	97	72	264
Hampshire . . .	53	38	38	129
Total . . .	148	135	110	393

If twice or thrice the number had been collected,
the three forms would probably have been found
nearly equal; I infer this from considering the above
figures, and from my son telling me that if he had

collected in another spot, he felt sure that the mid-styled plants would have been in excess. I several times sowed small parcels of seed, and raised all three forms; but I neglected to record the parent-form, excepting in one instance, in which I raised from short-styled seed twelve plants, of which only one turned out long-styled, four mid-styled, and seven short-styled.

Two plants of each form were protected from the access of insects during two successive years, and in the autumn they yielded very few capsules and presented a remarkable contrast with the adjoining uncovered plants, which were densely covered with capsules. In 1863 a protected long-styled plant produced only five poor capsules; two mid-styled plants produced together the same number; and two short-styled plants only a single one. These capsules contained very few seeds; yet the plants were fully productive when artificially fertilised under the net. In a state of nature the flowers are incessantly visited for their nectar by hive- and other bees, various Diptera and Lepidoptera.* The nectar is secreted all round the base of the ovarium; but a passage is formed along the upper and inner side of the flower by the lateral deflection (not repre-sented in the diagram) of the basal portions of the filaments; so that insects invariably alight on the pro-jecting stamens and pistil, and insert their proboscides along the upper and inner margin of the corolla. We can now see why the ends of the stamens with their anthers, and the ends of the pistils with their stigmas,

---

* H. Müller gives a list of the species, 'Die Befruchtung der Blumen,' p. 196. It appears that one bee, the *Cilissa melanura*, almost confines its visits to this plant.

are a little upturned, so that they may be brushed by
the lower hairy surfaces of the insects' bodies. The
shortest stamens which lie enclosed within the calyx of
the long- and mid-styled forms can be touched only by
the proboscis and narrow chin of a bee; hence they
have their ends more upturned, and they are graduated
in length, so as to fall into a narrow file, sure to be
raked by the thin intruding proboscis. The anthers of
the longer stamens stand laterally farther apart and are
more nearly on the same level, for they have to brush
against the whole breadth of the insect's body. In
very many other flowers the pistil, or the stamens, or
both, are rectangularly bent to one side of the flower.
This bending may be permanent, as with *Lythrum*
and many others, or may be effected, as in *Dictam-
nus fraxinella* and others, by a temporary movement,
which occurs in the case of the stamens when the
anthers dehisce, and in the case of the pistil when
the stigma is mature; but these two movements do
not always take place simultaneously in the same
flower. Now I have found no exception to the rule,
that when the stamens and pistil are bent, they bend
to that side of the flower which secretes nectar, even
though there be a rudimentary nectary of large size
on the opposite side, as in some species of *Corydalis*.
When nectar is secreted on all sides, they bend to
that side where the structure of the flower allows the
easiest access to it, as in *Lythrum*, various Papilio-
naceæ, and others. The rule consequently is, that
when the pistils and stamens are curved or bent, the
stigma and anthers are thus brought into the pathway leading to the nectary. There are a few cases
which seem to be exceptions to this rule, but they are
not so in truth; for instance, in the Gloriosa lily, the
stigma of the grotesque and rectangularly bent pistil

is brought, not into any pathway from the outside towards the nectar-secreting recesses of the flower, but into the circular route which insects follow in proceeding from one nectary to the other. In *Scrophularia aquatica* the pistil is bent downwards from the mouth of the corolla, but it thus strikes the pollen-dusted breast of the wasps which habitually visit these ill-scented flowers. In all these cases we see the supreme dominating power of insects on the structure of flowers, especially of those which have irregular corollas. Flowers which are fertilised by the wind must of course be excepted; but I do not know of a single instance of an irregular flower which is thus fertilised.

Another point deserves notice. In each of the three forms two sets of stamens correspond in length with the pistils in the other two forms. When bees suck the flowers, the anthers of the longest stamens, bearing the green pollen, are rubbed against the abdomen and the inner sides of the hind legs, as is likewise the stigma of the long-styled form. The anthers of the mid-length stamens and the stigma of the mid-styled form are rubbed against the under side of the thorax and between the front pair of legs. And, lastly, the anthers of the shortest stamens and the stigma of the short-styled form are rubbed against the proboscis and chin; for the bees in sucking the flowers insert only the front part of their heads into the flower. On catching bees, I observed much green pollen on the inner sides of the hind legs and on the abdomen, and much yellow pollen on the under side of the thorax. There was also pollen on the chin, and, it may be presumed, on the proboscis, but this was difficult to observe. I had, however, independent proof that pollen is carried on the proboscis; for a small branch of a protected short-styled plant (which produced spontaneously only two

capsules) was accidentally left during several days
pressing against the net, and bees were seen inserting
their proboscides through the meshes, and in consequence numerous capsules were formed on this one
small branch. From these several facts it follows that
insects will generally carry the pollen of each form from
the stamens to the pistil of corresponding length; and
we shall presently see the importance of this adaptation. It must not, however, be supposed that the bees
do not get more or less dusted all over with the several
kinds of pollen; for this could be seen to occur with
the green pollen from the longest stamens. Moreover
a case will presently be given of a long-styled plant
producing an abundance of capsules, though growing quite by itself, and the flowers must have been
fertilised by their own two kinds of pollen; but
these capsules contained a very poor average of seed.
Hence insects, and chiefly bees, act both as general
carriers of pollen, and as special carriers of the right
sort.

Wirtgen remarks* on the variability of this plant in
the branching of the stem, in the length of the bracteæ,
size of the petals, and in several other characters. The
plants which grew in my garden had their leaves,
which differed much in shape, arranged oppositely,
alternately, or in whorls of three. In this latter case
the stems were hexagonal; those of the other plants
being quadrangular. But we are concerned chiefly,
with the reproductive organs: the upward bending of
the pistil is variable, and especially in the short-styled
form, in which it is sometimes straight, sometimes
slightly curved, but generally bent at right angles.
The stigma of the long-styled pistil frequently has

---

* 'Verhand. des naturhist. Vereins, für Pr. Rheinl.' 5. Jahrgang,
1848, pp. 11, 13.

longer papillæ or is rougher than that of the mid-styled, and the latter than that of the short-styled; but this character, though fixed and uniform in the two forms of *Primula veris*, &c., is here variable, for I have seen mid-styled stigmas rougher than those of the long-styled.* The degree to which the longest and mid-length stamens are graduated in length and have their ends upturned is variable; sometimes all are equally long. The colour of the green pollen in the longest stamens is variable, being sometimes pale greenish-yellow; in one short-styled plant it was almost white. The grains vary a little in size: I examined one short-styled plant with the grains above the average size; and I have seen a long-styled plant with the grains from the mid-length and shortest anthers of the same size. We here see great variability in many important characters; and if any of these variations were of service to the plant, or were correlated with useful functional differences, the species is in that state in which natural selection might readily do much for its modification.

On the Power of Mutual Fertilisation between the three Forms.

Nothing shows more clearly the extraordinary complexity of the reproductive system of this plant, than the necessity of making eighteen distinct unions in order to ascertain the relative fertilising power of the

---

* The plants which I observed grew in my garden, and probably varied rather more than those growing in a state of nature. H. Müller has described the stigmas of all three forms with great care, and he appears to have found the stigmatic papillæ differing constantly in length and structure in the three forms, being longest in the long-styled form.

three forms. Thus the long-styled form has to be fertilised with pollen from its own two kinds of anthers,
from the two in the mid-styled, and from the two in
the short-styled form. The same process has to be
repeated with the mid-styled and short-styled forms.
It might have been thought sufficient to have tried on
each stigma the green pollen, for instance, from either
the mid- or short-styled longest stamens, and not
from both; but the result proves that this would
have been insufficient, and that it was necessary to
try all six kinds of pollen on each stigma. As in
fertilising flowers there will always be some failures,
it would have been advisable to have repeated each of
the eighteen unions a score of times; but the labour
would have been too great; as it was, I made 223
unions, i.e. on an average I fertilised above a dozen
flowers in the eighteen different methods. Each flower
was castrated; the adjoining buds had to be removed,
so that the flowers might be safely marked with
thread, wool, &c.; and after each fertilisation the stigma
was examined with a lens to see that there was sufficient pollen on it. Plants of all three forms were
protected during two years by large nets on a frame-
work; two plants were used during one or both years,
in order to avoid any individual peculiarity in a particular plant. As soon as the flowers had withered,
the nets were removed; and in the autumn the capsules were daily inspected and gathered, the ripe
seeds being counted under the microscope. I have
given these details that confidence may be placed
in the following tables, and as some excuse for two
blunders which, I believe, were made. These blunders
are referred to, with their probable cause, in two
foot-notes to the tables. The erroneous numbers, however, are entered in the tables, that it may not be sup-

posed that I have in any one instance tampered with the results.

A few words explanatory of the three tables must be given. Each is devoted to one of the three forms, and is divided into six compartments. The two upper ones in each table show the number of good seeds resulting from the application to the stigma of pollen from the two sets of stamens which correspond in length with the pistil of that form, and which are borne by the other two forms. Such unions are of a legitimate nature. The two next lower compartments show the result of the application of pollen from the two sets of stamens, not corresponding in length with the pistil, and which are borne by the other two forms. These unions are illegitimate. The two lowest compartments show the result of the application of each form's own two kinds of pollen from the two sets of stamens belonging to the same form, and which do not equal the pistil in length. These unions are likewise illegitimate. The term own-form pollen here used does not mean pollen from the flower to be fertilised—for this was never used—but from another flower on the same plant, or more commonly from a distinct plant of the same form. The figure (0) means that no capsule was produced, or if a capsule was produced that it contained no good seed. In some part of each row of figures in each compartment, a short horizontal line may be seen; the unions above this line were made in 1862, and below it in 1863. It is of importance to observe this, as it shows that the same general result was obtained during two successive years; but more especially because 1863 was a very hot and dry season, and the plants had occasionally to be watered. This did not prevent the full complement of seed being produced from the more fertile unions; but it rendered the less fertile ones even more sterile than they otherwise would have been. I have seen striking instances of this fact in making illegitimate and legitimate unions with Primula; and it is well known that the conditions of life must be highly favourable to give any chance of success in producing hybrids between species which are crossed with difficulty.

TABLE 23.—*Long-styled Form.*

I.		II.	
Legitimate union.		*Legitimate union.*	
13 flowers fertilised by the longest stamens of the mid-styled. These stamens equal in length the pistil of the long-styled.		13 flowers fertilised by the longest stamens of the short-styled. These stamens equal in length the pistil of the long-styled.	
Product of good seed in each capsule.		Product of good seed in each capsule.	
36	53	159	104
81	0	43	119
0	0	96 poor seed.	96
0	0	103	99
0	0	0	131
—	0	0	116
45		—	
41		114	
38 per cent. of these flowers yielded capsules. Each capsule contained, on an average, 51·2 seeds.		84 per cent. of these flowers yielded capsules. Each capsule contained, on an average, 107·3 seeds.	
III.		IV.	
Illegitimate union.		*Illegitimate union.*	
14 flowers fertilised by the shortest stamens of the mid-styled.		12 flowers fertilised by the mid-length stamens of the short-styled.	
3	0	20	0
0	0	0	0
0	0	0	0
0	0	0	0
0	0	—	0
—	0	0	0
0	0	0	
0			
Too sterile for any average.		Too sterile for any average.	

TABLE 23.—*Long-styled Form—continued.*

V. Illegitimate union. 15 flowers fertilised by own-form mid-length stamens.		VI. Illegitimate union. 15 flowers fertilised by own-form shortest stamens.	
2	—	4	—
10	0	8	0
23	0	4	0
0	0	0	0
0	0	0	0
0	0	0	0
0	0	0	0
0	0	0	0
Too sterile for any average.		Too sterile for any average.	

Besides the above experiments, I fertilised a considerable number of long-styled flowers with pollen, taken by a camel's-hair brush, from both the midlength and shortest stamens of their own form: only 5 capsules were produced, and these yielded on an average 14·5 seeds. In 1863 I tried a much better experiment: a long-styled plant was grown by itself, miles away from any other plant, so that the flowers could have received only their own two kinds of pollen. The flowers were incessantly visited by bees, and their stigmas must have received successive applications of pollen on the most favourable days and at the most favourable hours: all who have crossed plants know that this highly favours fertilisation. This plant produced an abundant crop of capsules; I took by chance 20 capsules, and these contained seeds in number as follows:—

20	20	35	21	19
26	24	12	23	10
7	30	27	29	13
20	12	29	19	35

This gives an average of 21·5 seeds per capsule. As we know that the long-styled form, when standing near plants of the other two forms and fertilised by insects, produces on an average 93 seeds per capsule, we see that this form, fertilised by its own two pollens, yields only between one-fourth and one-fifth of the full number of seed. I have spoken as if the plant had received both its own kinds of pollen, and this is, of course, possible; but, from the enclosed position of the shortest stamens, it is much more probable that the stigma received exclusively pollen from the mid-length stamens; and this, as may be seen in compartment V. in Table 23, is the more fertile of the two self-unions.

TABLE 24.—*Mid-styled Form.*

I.		II.	
Legitimate union.		*Legitimate union.*	
12 flowers fertilised by the mid-length stamens of the long-styled. These stamens equal in length the pistil of the mid-styled.		12 flowers fertilised by the mid-length stamens of the short-styled. These stamens equal in length the pistil of the mid-styled.	
Product of good seed in each capsule.		Product of good seed in each capsule.	
138	122	112	109
149	50	130	143
147	151	143	124
109	119	100	145
133	138	33	12
144	0	—	141
—		104	

92 per cent. of the flowers (probably 100 per cent.) yielded capsules. Each capsule contained, on an average, 127·3 seeds.

100 per cent. of the flowers yielded capsules. Each capsule contained, on an average, 108·0 seeds; or, excluding capsules with less than 20 seeds, the average is 116·7 seeds.

TABLE 24.—*Mid-styled Form—continued.*

III.			IV.	
Illegitimate union.			*Illegitimate union.*	
13 flowers fertilised by the shortest stamens of the long-styled.			15 flowers fertilised by the longest stamens of the short-styled.	
83	12		130	86
0	19		115	113
0	85 {seeds small and poor.		14	29
			6	17
—	0		2	113
44	0		9	79
44	0		—	128
45	0		132	0

54 per cent. of the flowers yielded capsules. Each. capsule contained, on an average, 47·4 seeds; or, excluding capsules with less than 20 seeds, the average is 60·2 seeds.

93 per cent. of the flowers yielded capsules. Each capsule contained, on an average, 69·5 seeds; or, excluding capsules with less than 20 seeds, the average is 102·8.

V.		VI.	
Illegitimate union.		*Illegitimate union.*	
12 flowers fertilised by own-form longest stamens.		12 flowers fertilised by own-form shortest stamens.	
92	0	0	0
9	0	0	0
63	0	0	0
—	0	—	0
136 ? *	0	0	0
0	0	0	0
0		0	

Excluding the capsule with 136 seeds, 25 per cent. of the flowers yielded capsules, and each capsule contained, on an average, 54·6 seeds; or, excluding capsules with less than 20 seeds, the average is 77·5.

Not one flower yielded a capsule.

* I have hardly a doubt that this result of 136 seeds in compartment V. was due to a gross error. The flowers to be fertilised by their own longest stamens were first marked by "white thread," and those by the mid-length stamens of the long-styled form by "white silk;" a flower fertilised in the later manner would have yielded about 136 seeds, and it may be observed that one such pod is missing, viz. at the bottom of compartment I. Therefore I have hardly any doubt that I fertilised a flower marked with "white thread" as if it had been marked with "white silk." With respect to the capsule which yielded 92 seeds, in the same column with that which yielded 136, I do not know what to think. I endeavoured to prevent pollen dropping from an upper to a lower

Besides the experiments in the above table, I fertilised a considerable number of mid-styled flowers with
pollen, taken by a camel's-hair brush, from both the
longest and shortest stamens of their own form : only
5 capsules were produced, and these yielded on an
average 11·0 seeds.

TABLE 25.—*Short-styled Form.*

I. Legitimate union. 12 flowers fertilised by the shortest stamens of the long-styled. These stamens equal in length the pistil of the short-styled.		II. Legitimate union. 13 flowers fertilised by the shortest stamens of the mid-styled. These stamens equal in length the pistil of the short-styled.	
69	56	93	69
61	88	77	69
88	112	48	53
66	111	43	9
0	62	0	0
0	100	0	0
—		—	0
83 per cent. of the flowers yielded capsules. Each capsule contained, on an average, 81·3 seeds.		61 per cent. of the flowers yielded capsules. Each capsule contained, on an average, 64·6 seeds.	

III. Illegitimate union. 10 flowers fertilised by the mid-length stamens of the long-styled.		IV. Illegitimate union. 10 flowers fertilised by the longest stamens of the mid-styled.	
0	14	0	0
0	0	0	0
0	0	0	0
0	0	0	0
—	0	—	0
23		0	
Too sterile for any average.		Too sterile for any average.	

flower, and I tried to remember to
wipe the pincers carefully after
each fertilisation ; but in making
eighteen different unions, sometimes on windy days, and pestered
by bees and flies buzzing about,
some few errors could hardly be
avoided. One day I had to keep
a third man by me all the time to
prevent the bees visiting the uncovered plants, for in a few
seconds' time they might have
done irreparable mischief. It was
also extremely difficult to exclude
minute Diptera from the net. In
1862 I made the great mistake of
placing a mid-styled and long-
styled under the same huge net :
in 1863 I avoided this error.

TABLE 25.—*Short-styled Form—continued.*

V. *Illegitimate union.* 10 flowers fertilised by own-form longest stamens.		VI. *Illegitimate union.* 10 flowers fertilised by own-form mid-length stamens.	
0	0	64 ? *	0
0	0	0	0
0	0	0	0
—	0	—	0
0	0	21	0
0		9	
Too sterile for any average.		Too sterile for any average.	

Besides the experiments in the table, I fertilised a number of flowers without particular care with their own two kinds of pollen, but they did not produce a single capsule.

Summary of the Results.

Long-styled form.—Twenty-six flowers fertilised legitimately by the stamens of corresponding length, borne by the mid- and short-styled forms, yielded 61·5 per cent. of capsules, which contained on an average 89·7 seeds.

Twenty-six long-styled flowers fertilised illegitimately by the other stamens of the mid- and short-styled forms yielded only two very poor capsules.

Thirty long-styled flowers fertilised illegitimately by their own-form two sets of stamens yielded only eight very poor capsules; but long-styled flowers fertilised

---

* I suspect that by mistake I fertilised this flower in compartment VI. with pollen from the shortest stamens of the long-styled form, and it would then have yielded about 64 seeds. Flowers to be thus fertilised were marked with black silk; those with pollen from the mid-length stamens of the short-styled with black thread; and thus probably the mistake arose.

by bees with pollen from their own stamens produced numerous capsules containing on an average 21·5 seeds.

Mid-styled form.—Twenty-four flowers legitimately fertilised by the stamens of corresponding length, borne by the long and short-styled forms, yielded 96 (probably 100) per cent. of capsules, which contained (excluding one capsule with 12 seeds) on an average 117·2 seeds.

Fifteen mid-styled flowers fertilised illegitimately by the longest stamens of the short-styled form yielded 93 per cent. of capsules, which (excluding four capsules with less than 20 seeds) contained on an average 102·8 seeds.

Thirteen mid-styled flowers fertilised illegitimately by the mid-length stamens of the long-styled form yielded 54 per cent. of capsules, which (excluding one with 19 seeds) contained on an average 60·2 seeds.

Twelve mid-styled flowers fertilised illegitimately by their own-form longest stamens yielded 25 per cent. of capsules, which (excluding one with 9 seeds) contained on an average 77·5 seeds.

Twelve mid-styled flowers fertilised illegitimately by their own-form shortest stamens yielded not a single capsule.

Short-styled form.—Twenty-five flowers fertilised legitimately by the stamens of corresponding length, borne by the long and mid-styled forms, yielded 72 per cent. of capsules, which (excluding one capsule with only 9 seeds) contained on an average 70·8 seeds.

Twenty short-styled flowers fertilised illegitimately by the other stamens of the long and mid-styled forms yielded only two very poor capsules.

Twenty short-styled flowers fertilised illegitimately

by their own stamens yielded only two poor (or perhaps three) capsules.

If we take all six legitimate unions together, and
all twelve illegitimate unions together, we get the following results :—

TABLE 26.

Nature of Union.	Number of Flowers fertilised.	Number of Capsules produced.	Average Number of Seeds per Capsule.	Average Number of Seeds per Flower fertilised.
The six legitimate unions	75	56	96·29	71·89
The twelve illegitimate unions . .	146	36	44·72	11·03

Therefore the fertility of the legitimate unions to that
of the illegitimate, as judged by the proportion of the
fertilised flowers which yielded capsules, is as 100 to
33 ; and judged by the average number of seeds per
capsule, as 100 to 46.

From this summary and the several foregoing tables
we see that it is only pollen from the longest stamens
which can fully fertilise the longest pistil; only that
from the mid-length stamens, the mid-length pistil;
and only that from the shortest stamens, the shortest
pistil. And now we can comprehend the meaning of
the almost exact correspondence in length between
the pistil in each form and a set of six stamens
in two of the other forms ; for the stigma of each
form is thus rubbed against that part of the insect's
body which becomes charged with the proper pollen.
It is also evident that the stigma of each form,
fertilised in three different ways with pollen from
the longest, mid-length, and shortest stamens, is acted
on very differently, and conversely that the pollen from

the twelve longest, twelve mid-length, and twelve
shortest stamens acts very differently on each of the
three stigmas; so that there are three sets of female
and of male organs. Moreover, in most cases the six
stamens of each set differ somewhat in their fertilising
power from the six corresponding ones in one of the
other forms. We may further draw the remarkable
conclusion that the greater the inequality in length
between the pistil and the set of stamens, the pollen
of which is employed for its fertilisation, by so much
is the sterility of the union increased. There are no
exceptions to this rule. To understand what follows
the reader should look to Tables 23, 24, and 25, and
to the diagram Fig. 10, p. 139. In the long-styled form
the shortest stamens obviously differ in length from
the pistil to a greater degree than do the mid-length
stamens; and the capsules produced by the use of
pollen from the shortest stamens contain fewer seeds
than those produced by the pollen from the mid-
length stamens. The same result follows with the
long-styled form, from the use of the pollen of the
shortest stamens of the mid-styled form and of the
mid-length stamens of the short-styled form. The
same rule also holds good with the mid-styled and
short-styled forms, when illegitimately fertilised with
pollen from the stamens more or less unequal in
length to their pistils. Certainly the difference in
sterility in these several cases is slight; but, as far as
we are enabled to judge, it always increases with the
increasing inequality of length between the pistil and
the stamens which are used in each case.

The correspondence in length between the pistil in
each form and a set of stamens in the other two forms,
is probably the direct result of adaptation, as it is of
high service to the species by leading to full and

legitimate fertilisation. But the rule of the increased sterility of the illegitimate unions according to the greater inequality in length between the pistils and stamens employed for the union can be of no service. With some heterostyled dimorphic plants the difference of fertility between the two illegitimate unions appears at first sight to be related to the facility of self-fertilisation; so that when from the position of the parts the liability in one form to self-fertilisation is greater than in the other, a union of this kind has been checked by having been rendered the more sterile of the two. But this explanation does not apply to Lythrum; thus the stigma of the long-styled form is more liable to be illegitimately fertilised with pollen from its own mid-length stamens, or with pollen from the mid-length stamens of the short-styled form, than by its own shortest stamens or those of the mid-styled form; yet the two former unions, which it might have been expected would have been guarded against by increased sterility, are much less sterile than the other two unions which are much less likely to be 'effected. The same relation holds good even in a more striking manner with the mid-styled form, and with the short-styled form as far as the extreme sterility of all its illegitimate unions allows of any comparison. We are led, therefore, to conclude that the rule of increased sterility in accordance with increased inequality in length between the pistils and stamens, is a purposeless result, incidental on those changes through which the species has passed in acquiring certain characters fitted to ensure the legitimate fertilisation of the three forms.

Another conclusion which may be drawn from Tables 23, 24, and 25, even from a glance at them,

M

is that the mid-styled form differs from both the others in its much higher capacity for fertilisation in various ways. Not only did the twenty-four flowers legitimately fertilised by the stamens of corresponding lengths, all, or all but one, yield capsules rich in seed; but of the other four illegitimate unions, that by the longest stamens of the short-styled form was highly fertile, though less so than the two legitimate unions, and that by the mid-length stamens of the long-styled form was fertile to a considerable degree; the remaining two illegitimate unions, namely, with this form's own pollen, were sterile, but in different degrees. So that the mid-styled form, when fertilised in the six different possible methods, evinces five grades of fertility. By comparing compartments III. and VI. in Table 24 we may see that the action of the pollen from the shortest stamens of the long-styled and mid-styled forms is widely different; in the one case above half the fertilised flowers yielded capsules containing a fair number of seeds; in the other case not one capsule was produced. So, again, the green, large-grained pollen from the longest stamens of the short-styled and mid-styled forms (in compartments IV. and V.) is widely different. In both these cases the difference in action is so plain that it cannot be mistaken, but it can be corroborated. If we look to Table 25 to the legitimate action of the shortest stamens of the long- and mid-styled forms on the short-styled form, we again see a similar but slighter difference, the pollen of the shortest stamens of the mid-styled form yielding a smaller average of seed during the two years of 1862 and 1863 than that from the shortest stamens of the long-styled form. Again, if we look to Table 23, to the legitimate action on the long-styled form of the green pollen of the two

sets of longest stamens, we shall find exactly the same
result, viz. that the pollen from the longest stamens of
the mid-styled form yielded during both years fewer
seeds than that from the longest stamens of the
short-styled form.   Hence it is certain that the two
kinds of pollen produced by the mid-styled form are
less potent than the two similar kinds of pollen produced by the corresponding stamens of the other two
forms.

In close connection with the lesser potency of the
two kinds of pollen of the mid-styled form is the fact
that, according to H. Müller, the grains of both are
a little less in diameter than the corresponding grains
produced by the other two forms.   Thus the grains
from the longest stamens of the mid-styled form are
9 to 10, whilst those from the corresponding stamens
of the short-styled form are $9\frac{1}{2}$ to $10\frac{1}{2}$ in diameter.
So, again, the grains from the shortest stamens of the
mid-styled are 6, whilst those from the corresponding
stamens of the long-styled are 6 to $6\frac{1}{2}$ in diameter.
It would thus appear as if the male organs of the
mid-styled form, though not as yet rudimentary, were
tending in this direction.   On the other hand, the
female organs of this form are in an eminently efficient
state, for the naturally fertilised capsules yielded a
considerably larger average number of seeds than
those of the other two forms—almost every flower
which was artificially fertilised in a legitimate manner
produced a capsule—and most of the illegitimate
unions were highly productive.   The mid-styled form
thus appears to be highly feminine in nature; and although, as just remarked, it is impossible to consider
its two well-developed sets of stamens which produce
an abundance of pollen as being in a rudimentary
condition, yet we can hardly avoid connecting as

balanced the higher efficiency of the female organs in this form with the lesser efficiency and lesser size of its two kinds of pollen-grains. The whole case appears to me a very curious one.

It may be observed in Tables 23 to 25 that some of the illegitimate unions yielded during neither year a single seed; but, judging from the long-styled plants, it is probable, if such unions were to be effected repeatedly by the aid of insects under the most favourable conditions, some few seeds would be produced in every case. Anyhow, it is certain that in all twelve illegitimate unions the pollen-tubes penetrated the stigma in the course of eighteen hours. At first I thought that two kinds of pollen placed together on the same stigma would perhaps yield more seed than one kind by itself; but we have seen that this is not so with each form's own two kinds of pollen; nor is it probable in any case, as I occasionally got, by the use of a single kind of pollen, fully as many seeds as a capsule naturally fertilised ever produces. Moreover the pollen from a single anther is far more than sufficient to fertilise fully a stigma; hence, in this as with so many other plants, more than twelve times as much of each kind of pollen is produced as is necessary to ensure the full fertilisation of each form. From the dusted condition of the bodies of the bees which I caught on the flowers, it is probable that pollen of various kinds is often deposited on all three stigmas; but from the facts already given with respect to the two forms of Primula, there can hardly be a doubt that pollen from the stamens of corresponding length placed on a stigma would be prepotent over any other kind of pollen and obliterate its effects, —even if the latter had been placed on the stigma some hours previously.

Finally, it has now been shown that *Lythrum salicaria* presents the extraordinary case of the same species bearing three females, different in structure and function, and three or even five sets (if minor differences are considered) of males; each set consisting of half-a-dozen, which likewise differ from one another in structure and function.

Lythrum Græfferi.—I have examined numerous dried flowers of this species, each from a separate plant, sent me from Kew. Like *L. salicaria*, it is trimorphic, and the three forms apparently occur in about equal numbers. In the long-styled form the pistil projects about one-third of the length of the calyx beyond its mouth, and is therefore relatively much shorter than in *L. salicaria*; the globose and hirsute stigma is larger than that of the other two forms; the six mid-length stamens, which are graduated in length, have their anthers standing close above and close beneath the mouth of the calyx; the six shortest stamens rise rather above the middle of the calyx. In the mid-styled form the stigma projects just above the mouth of the calyx, and stands almost on a level with the mid-length stamens of the long and short-styled forms; its own longest stamens project well above the mouth of the calyx, and stand a little above the level of the stigma of the long-styled form. In short, without entering on further details, there is a close general correspondence in structure between this species and *L. salicaria*, but with some differences in the proportional lengths of the parts. The fact of each of the three pistils having two sets of stamens of corresponding lengths, borne by the two other forms, comes out conspicuously. In the mid-styled form the pollen-grains from the longest stamens are nearly double the diameter of those from the shortest stamens; so that there is a greater difference in this respect than in *L. salicaria*. In the long-styled form, also, the difference in diameter between the pollen-grains of the mid-length and shortest stamens is greater than in *L. salicaria*. These comparisons, however, must be received with caution, as they were made on specimens soaked in water after having been long kept dry.

Lythrum thymifolia.—This form, according to Vaucher,[*] is

* 'Hist. Phys. des Plantes d'Europe,' tom. ii. (1841), pp. 369, 371.

dimorphic, like Primula, and therefore presents only two forms.
I received two dried flowers from Kew, which consisted of the
two forms; in one the stigma projected far beyond the calyx, in
the other it was included within the calyx; in this latter form
the style was only one-fourth of the length of that in the other
form.   There are only six stamens; these are somewhat graduated in length, and their anthers in the short-styled form stand
a little above the stigma, but yet by no means equal in length
the pistil of the long-styled form.   In the latter the stamens
are rather shorter than those in the other form.   The six
stamens alternate with the petals, and therefore correspond
homologically with the longest stamens of *L. salicaria* and *L.
Græfferi.*

Lythrum hyssopifolia.—This species is said by Vaucher, but I
believe erroneously, to be dimorphic.   I have examined dried
flowers from twenty-two separate plants from various localities,
sent to me by Mr. Hewett C. Watson, Professor Babington, and
others.   These were all essentially alike, so that the species
cannot be heterostyled.   The pistil varies somewhat in length,
but when unusually long, the stamens are likewise generally
long; in the bud the stamens are short; and Vaucher was
perhaps thus deceived.   There are from six to nine stamens,
graduated in length.   The three stamens, which vary in being
either present or absent, correspond with the six shorter stamens
of *L. salicaria* and with the six which are always absent in *L.
thymifolia.* The stigma is included within the calyx, and stands
in the midst of the anthers, and would generally be fertilised
by them; but as the stigma and anthers are upturned, and as,
according to Vaucher, there is a passage left in the upper side
of the flower to the nectary, there can hardly be a doubt that
the flowers are visited by insects, and would occasionally be
cross-fertilised by them, as surely as the flowers of the short-
styled *L. salicaria,* the pistil of which and the corresponding
stamens in the other two forms closely resemble those of *L. hyssopifolia.* According to Vaucher and Lecoq,* this species, which
is an annual, generally grows almost solitarily, whereas the
three preceding species are social; and this fact alone would
almost have convinced me that *L. hyssopifolia* was not heterostyled, as such plants cannot habitually live isolated any better
than one sex of a diœcious species.

---

* 'Géograph. Bot. de l'Europe,' tom. vi. 1857, p. 157.

We thus see that within this genus some species are hetero-styled and trimorphic; one apparently heterostyled and dimorphic, and one homostyled.

Nesæa verticillata.—I raised a number of plants from seed sent me by Professor Asa Gray, and they presented three forms. These differed from one another in the proportional lengths of their organs of fructification and in all respects, in very nearly the same way as the three forms of *Lythrum Græfferi*. The green pollen-grains from the longest stamens, measured along their longer axis and not distended with water, were $\frac{13}{7000}$ of an inch in length; those from the mid-length stamens $\frac{9\text{-}10}{7000}$, and those from the shortest stamens $\frac{8\text{-}9}{7000}$ of an inch. So that the largest pollen-grains are to the smallest in diameter as 100 to 65. This plant inhabits swampy ground in the United States. According to Fritz Müller,[*] a species of this genus in St. Catharina, in Southern Brazil, is homostyled.

Lagerstrœmia Indica.—This plant, a member of the Lythraceæ, may perhaps be heterostyled, or may formerly have been so. It is remarkable from the extreme variability of its stamens. On a plant, growing in my hothouse, the flowers included from nineteen to twenty-nine short stamens with yellow pollen, which correspond in position with the shortest stamens of Lythrum; and from one to five (the latter number being the commonest) very long stamens, with thick flesh-coloured filaments and green pollen, corresponding in position with the longest stamens of Lythrum. In one flower, two of the long stamens produced green, while a third produced yellow pollen, although the filaments of all three were thick and flesh-coloured. In an anther of another flower, one cell contained green and the other yellow pollen. The green and yellow pollen-grains from the stamens of different length are of the same size. The pistil is a little bowed upwards, with the stigma seated between the anthers of the short and long stamens, so that this plant was mid-styled. Eight flowers were fertilised with green pollen, and six with yellow pollen, but not one set fruit. This latter fact by no means proves that the plant is hetero-styled, as it may belong to the class of self-sterile species. Another plant growing in the Botanic Gardens at Calcutta, as Mr. J. Scott informs me, was long-styled, and it was èqually

---

[*] 'Bot. Zeitung,' 1868, p. 112.

sterile with its own pollen; whilst a long-styled plant of
L. reginæ, though growing by itself, produced fruit. I examined
dried flowers from two plants of *L. parviflora*, both of which
were long-styled, and they differed from *L. Indica* in having
eight long stamens with thick filaments, and a crowd of shorter
stamens.   Thus the evidence whether *L. Indica* is heterostyled is curiously conflicting: the unequal number of the short
and long stamens, their extreme variability, and especially the
fact of their pollen-grains not differing in size, are strongly
opposed to this belief; on the other hand, the difference in
length of the pistils in two of the plants, their sterility with
their own pollen, and the difference in length and structure of
the two sets of stamens in the same flower, and in the colour of
their pollen, favour the belief. We know that when plants of
any kind revert to a former condition, they are apt to be highly
variable, and the two halves of the same organ sometimes differ
much, as in the case of the above-described anther of the
Lagerstrœmia; we may therefore suspect that this species was
once heterostyled, and that it still retains traces of its former
state, together with a tendency to revert more completely to it.
It deserves notice, as bearing on the nature of Lagerstrœmia,
that in *Lythrum hyssopifolia*, which is a homostyled species, some
of the shorter stamens vary in being either present or absent;
and that these same stamens are altogether absent in *L. thymifolia*. In another genus of the Lythraceæ, namely Cuphea, three
species raised by me from seed certainly were homostyled;
nevertheless their stamens consisted of two sets, differing in
length and in the colour and thickness of their filaments, but
not in the size or colour of their pollen-grains; so that they
thus far resembled the stamens of Lagerstrœmia. I found that
Cuphea purpurea was highly fertile with its own pollen when
artificially aided, but sterile when insects were excluded.*

---

* Mr. Spence informs me that
in several species of the genus
Mollia (Tiliaceæ) which he collected in South America, the
stamens of the five outer cohorts
have purplish filaments and green
pollen, whilst the stamens of the
five inner cohorts have yellow
pollen. He therefore suspected
that these species might prove
to be heterostyled and trimorphic: but he did not notice the
length of the pistils.   In the
allied Luhea the outer purplish
stamens are destitute of anthers.
I procured some specimens of
Mollia lepidota and *speciosa* from
Kew, but could not make out that
their pistils differed in length
in different plants; and in all
those which I examined the
stigma stood close beneath the

Oxalis (Geraniaceæ).

In 1863 Mr. Roland Trimen wrote to me from the
Cape of Good Hope that he had there found species of
Oxalis which presented three forms; and of these he
enclosed drawings and dried specimens. Of one species
he collected 43 flowers from distinct plants, and they
consisted of 10 long-styled, 12 mid-styled, and 21
short-styled. Of another species he collected 13 flowers,
consisting of 3 long-styled, 7 mid-styled, and 3 short-
styled. In 1866 Prof. Hildebrand proved* by an examination of the specimens in several herbaria that 20
species are certainly heterostyled and trimorphic, and
51 others almost certainly so. He also made some interesting observations on living plants belonging to
one form alone; for at that time he did not possess
the three forms of any living species. During the
years 1864 to 1868 I occasionally experimented on
Oxalis speciosa, but until now have never found time
to publish the results. In 1871 Hildebrand published
an admirable paper† in which he shows in the case of
two species of Oxalis, that the sexual relations of the
three forms are nearly the same as in *Lythrum salicaria*. I will now give an abstract of his observations, and afterwards of my own less complete ones.
I may premise that in all the species seen by me, the
stigmas of the five straight pistils of the long-styled
form stand on a level with the anthers of the longest
stamens in the two other forms. In the mid-styled

---

uppermost anthers. The numerous
stamens are graduated in length,
and the pollen-grains from the
longest and shortest ones did not
present any marked difference in
diameter. Therefore these species
do not appear to be heterostyled.

* 'Monatsber. der Akad. der
Wiss. Berlin,' 1866, pp. 352, 372.
He gives drawings of the three
forms at p. 42 of his 'Geschlechter-
Vertheilung,' &c., 1867.

† 'Bot. Zeitung,' 1871, pp. 416
and 432.

form, the stigmas pass out between the filaments of the longest stamens (as in the short-styled form of Linum); and they stand rather nearer to the upper anthers than to the lower ones. In the short-styled

Fig. 11.

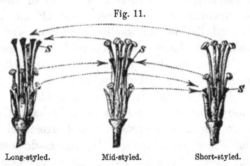

Long-styled.   Mid-styled.   Short-styled.

OXALIS SPECIOSA (with the petals removed).

S S S, stigmas. The dotted lines with arrows show which pollen must be carried to the stigmas for legitimate fertilisation.

form, the stigmas also pass out between the filaments nearly on a level with the tips of the sepals. The anthers in this latter form and in the mid-styled rise to the same height as the corresponding stigmas in the other two forms.

Oxalis Valdiviana.—This species, an inhabitant of the west coast of South America, bears yellow flowers. Hildebrand states that the stigmas of the three forms do not differ in any marked manner, but that the pistil of the short-styled form alone is destitute of hairs. The diameters of the pollen-grains are as follows :—

	Divisions of the Micrometer.
From the longest stamens of short-styled	8 to 9
„ mid-length „ „	7 „ 8
„ longest stamens of mid-styled	8
„ shortest „ „	6
„ mid-length stamens of long-styled . . .	7
„ shortest „ „	6

Therefore the extreme difference in diameter is as 8·5 to 6, or as 100 to 71. The results of Hildebrand's experiments are given in the following table, drawn up in accordance with my usual plan. He fertilised each form with pollen from the two sets of anthers of the same flower, and likewise from flowers on distinct plants belonging to the same form; but the effects of these two closely allied kinds of fertilisation differ so little that I have not kept them distinct.

TABLE 27.

Oxalis Valdiviana (from Hildebrand).

Nature of Union.	Number of Flowers fertilised.	Number of Capsules produced.	Number of Seeds per Capsule.
Long-styled form, by pollen of longest stamens of short-styled. Legitimate union	28	28	11·9
Long-styled form, by pollen of longest stamens of mid-styled. Legitimate union	21	21	12·0
Long-styled form, by pollen of own and own-form mid-length stamens. Illegitimate union	40	2	5·5
Long-styled form, by pollen of own and own-form shortest stamens. Illegitimate union	26	0	0
Long-styled form, by pollen of shortest stamens of short-styled. Illegitimate union	16	1	1
Long-styled form, by pollen of shortest stamens of mid-styled. Illegitimate union	9	0	0

Table 27—*continued.*

Oxalis Valdiviana (from Hildebrand).

Nature of Union.	Number of Flowers fertilised.	Number of Capsules produced.	Number of Seeds per Capsule.
Mid-styled form, by pollen of mid-length stamens of long-styled. Legitimate union.	38	38	11·3
Mid-styled form, by pollen of mid-length stamens of short-styled. Legitimate union	23	23	10·4
Mid-styled form, by pollen of own and own-form longest stamens. Illegitimate union	52	0	0
Mid-styled form, by pollen of own and own-form shortest stamens. Illegitimate union	30	1	6
Mid-styled form, by pollen of shortest stamens of long-styled. Illegitimate union	16	0	0
Mid-styled form, by pollen of longest stamens of short-styled. Illegitimate union	16	2	2·5
Short-styled form, by pollen of shortest stamens of long-styled. Legitimate union.	18	18	11·0
Short-styled form, by pollen of shortest stamens of mid-styled. Legitimate union .	10	10	11·3
Short-styled form, by pollen of own and own-form longest stamens. Illegitimate union	21	0	0
Short-styled form, by pollen of own and own-form mid-length stamens. Illegitimate union	22	0	0
Short-styled form, by pollen of longest stamens of mid-styled. Illegitimate union	4	0	0
Short-styled form, by pollen of mid-length stamens of long-styled. Illegitimate union	3	0	0

We here have the remarkable result that every one of 138 legitimately fertilised flowers on the three forms yielded capsules, containing on an average 11·33 seeds. Whilst of the 255 illegitimately fertilised flowers, only 6 yielded capsules, which contained 3·83 seeds on an average. Therefore the fertility of the six legitimate to that of the twelve illegitimate unions, as judged by the proportion of flowers that yielded capsules, is as 100 to 2, and as judged by the average number of seeds per capsule as 100 to 34. It may be added that some plants which were protected by nets did not spontaneously produce any fruit; nor did one which was left uncovered by itself and was visited by bees. On the other hand, scarcely a single flower on some uncovered plants of the three forms growing near together failed to produce fruit.

Oxalis Regnelli.—This species bears white flowers and inhabits Southern Brazil. Hildebrand says that the stigma of the long-styled form is somewhat larger than that of the mid-styled, and this than that of the short-styled. The pistil of the latter is clothed with a few hairs, whilst it is very hairy in the other two forms. The diameter of the pollen-grains from both sets of the longest stamens equals 9 divisions of the micrometer,—that from the mid-length stamens of the long-styled form between 8 and 9, and of the short-styled 8,—and that from the shortest stamens of both sets 7. So that the extreme difference in diameter is as 9 to 7 or as 100 to 78. The experiments made by Hildebrand, which are not so numerous as in the last case, are given in Table 28 in the same manner as before.

The results are nearly the same as in the last case, but more striking; for 41 flowers belonging to the three forms fertilised legitimately all yielded capsules,

TABLE 28.

Oxalis Regnelli (from Hildebrand).

Nature of Union.	Number of Flowers fertilised	Number of Capsules produced.	Average Number of Seeds per Capsule.
Long-styled form, by pollen of longest stamens of short-styled. Legitimate union	6	6	10·1
Long-styled form, by pollen of longest stamens of mid-styled. Legitimate union.	5	5	10·6
Long-styled form, by pollen of own mid-length stamens. Illegitimate union .	4	0	0
Long-styled form, by pollen of own shortest stamen. Illegitimate union. . .	1	0	0
Mid-styled form, by pollen of mid-length stamens of short-styled. Legitimate union	9	9	10·4
Mid-styled form, by pollen of mid-length stamens of long-styled. Legitimate union.	10	10	10·1
Mid-styled form, by pollen of own longest stamens. Illegitimate union . . .	9	0	0
Mid-styled form, by pollen of own shortest stamens. Illegitimate union . . .	2	0	0
Mid-styled form, by pollen of longest stamens of short-styled. Illegitimate union	1	0	0
Short-styled form, by pollen of shortest stamens of mid-styled. Legitimate union.	9	9	10·6
Short-styled form, by pollen of shortest stamens of long-styled. Legitimate union.	2	2	9·5
Short-styled form, by pollen of own mid-length stamens. Illegitimate union .	12	0	0
Short-styled form, by pollen of own longest stamens. Illegitimate union . .	9	0	0
Short-styled form, by pollen of mid-length stamens of long-styled. Illegitimate union	1	0	0

containing on an average 10·31 seeds; whilst 39
flowers fertilised illegitimately did not yield a single
capsule or seed.   Therefore the fertility of the six
legitimate to that of the several illegitimate unions,
as judged both by the proportion of flowers which
yielded capsules and by the average number of contained seeds, is as 100 to 0.

Oxalis speciosa.—This species, which bears pink
flowers, was introduced from the Cape of Good Hope.
A sketch of the reproductive organs of the three
forms (Fig. 11) has already been given.   The stigma
of the long-styled form (with the papillæ on its surface included) is twice as large as that of the short-
styled, and that of the mid-styled intermediate in size.
The pollen-grains from the stamens in the three forms
are in their longer diameters as follows :—

	Divisions of the Micrometer.
From the longest stamens of short-styled . . .	15 to 16
„ mid-length „ „ . . .	12 „ 13
„ longest stamens of mid-styled . . .	16
„ shortest „ „ . . .	11 to 12
„ mid-length stamens of long-styled . .	14
„ shortest „ „ . . .	12

Therefore the extreme difference in diameter is as
16 to 11, or as 100 to 69; but as the measurements
were taken at different times, they are probably only
approximately accurate.   The results of my experiments
in fertilising the three forms are given in the following
table.

TABLE 29.

Oxalis speciosa.

Nature of Union.	Number of Flowers fertilised.	Number of Capsules produced.	Average Number of Seeds per Capsule.
Long-styled form, by pollen of longest stamens of short-styled. Legitimate union	19	15	57·4
Long-styled form, by pollen of longest stamens of mid-styled. Legitimate union	4	3	59·0
Long-styled form, by pollen of own-form mid-length stamens. Illegitimate union	9	2	42·5
Long-styled form, by pollen of own-form shortest stamens. Illegitimate union .	11	0	0
Long-styled form, by pollen of shortest stamens of mid-styled. Illegitimate union	4	0	0
Long-styled form, by pollen of mid-length stamens of short-styled. Illegitimate union	12	5	30·0
Mid-styled form, by pollen of mid-length stamens of long-styled. Legitimate union	3	3	63·6
Mid-styled form, by pollen of mid-length stamens of short-styled. Legitimate union	4	4	56·3
Mid-styled form, by mixed pollen from both own-form longest and shortest stamens. Illegitimate union . . .	9	2	19
Mid-styled form, by pollen of longest stamens of short-styled. Illegitimate union	12	1	8

TABLE 29—*continued.*

Oxalis speciosa.

Nature of Union.	Number of Flowers fertilised.	Number of Capsules produced.	Average Number of Seeds per Capsule.
Short-styled form, by pollen of shortest stamens of mid - styled. Legitimate union	3	2	67
Short-styled form, by pollen of shortest stamens of long - styled. Legitimate union	3	3	54·3
Short-styled form, by pollen of own-form longest stamens. Illegitimate union .	5	1	8
Short-styled form, by pollen of own-form mid-length stamens. Illegitimate union	3	0	0
Short-styled form, by both pollens mixed together, of own-form longest and mid-length stamens. Illegitimate union .	13	0	0
Short-styled form, by pollen of longest stamens of mid - styled. Illegitimate union	7	0	0
Short-styled form, by pollen of mid-length stamens of long - styled. Illegitimate union	10	1	54

We here see that thirty-six flowers on the three forms legitimately fertilised yielded 30 capsules, these containing on an average 58·36 seeds. Ninety-five flowers illegitimately fertilised yielded 12 capsules, containing on an average 28·58 seeds. Therefore the fertility of the six legitimate to that of the twelve illegitimate unions, as judged by the proportion of flowers which yielded capsules, is as 100 to 15, and judged by the average number of seeds per capsule as 100 to 49. This plant, in comparison with the two South American species previously described, produces

N

many more seeds, and the illegitimately fertilised flowers are not quite so sterile.

Oxalis rosea.—Hildebrand possessed in a living state only the long-styled form of this trimorphic Chilian species.* The pollen-grains from the two sets of anthers differ in diameter as 9 to 7·5, or as 100 to 83. He has further shown that there is an analogous difference between the grains from the two sets of anthers of the same flower in five other species of Oxalis, besides those already described. The present species differs remarkably from the long-styled form of the three species previously experimented on, in a much larger proportion of the flowers setting capsules when fertilised with their own-form pollen. Hildebrand fertilised 60 flowers with pollen from the mid-length stamens (of either the same or another flower), and they yielded no less than 55 capsules, or 92 per cent. These capsules contained on an average 5·62 seeds; but we have no means of judging how near an approach this average makes to that from flowers legitimately fertilised. He also fertilised 45 flowers with pollen from the shortest stamens, and these yielded only 17 capsules, or 31 per cent., containing on an average only 2·65 seeds. We thus see that about thrice as many flowers, when fertilised with pollen from the mid-length stamens, produced capsules, and these contained twice as many seeds, as did the flowers fertilised with pollen from the shortest stamens. It thus appears (and we find some evidence of the same fact with *O. speciosa*), that the same rule holds good with Oxalis as with *Lythrum salicaria*; namely, that in any two unions, the greater the inequality in length between the pistils and stamens, or,

---

* 'Monatsber. der Akad. der Wiss. Berlin,' 1866, p. 372.

which is the same thing, the greater the distance of
the stigma from the anthers, the pollen of which is
used for fertilisation, the less fertile is the union,—
whether judged by the proportion of flowers which
set capsules, or by the average number of seeds per
capsule. The rule cannot be explained in this case
any more than in that of Lythrum, by supposing
that wherever there is greater liability to self-fertilisation, this is checked by the union being rendered more
sterile; for exactly the reverse occurs, the liability to
self-fertilisation being greatest in the unions between
the pistils and stamens which approach each other the
nearest, and these are the more fertile. I may add
that I also possessed some long-styled plants of this
species: one was covered by a net, and it set spontaneously a few capsules, though extremely few compared with those produced by a plant growing by
itself, but exposed to the visits of bees.

With most of the species of Oxalis the short-styled
form seems to be the most sterile of the three forms,
when these are illegitimately fertilised; and I will add
two other cases to those already given. I fertilised
29 short-styled flowers of *O. compressa* with pollen from
their own two sets of stamens (the pollen-grains of
which differ in diameter as 100 and 83), and not one
produced a capsule. I formerly cultivated during
several years the short-styled form of a species purchased under the name of *O. Bowii* (but I have some
doubts whether it was rightly named), and fertilised
many flowers with their own two kinds of pollen,
which differ in diameter in the usual manner, but
never got a single seed. On the other hand, Hildebrand says that the short-styled form of *O. Deppei*,
growing by itself, yields plenty of seed; but it is not
positively known that this species is heterostyled; and

N 2

the pollen-grains from the two sets of anthers do not differ in diameter.

Some facts communicated to me by Fritz Müller afford excellent evidence of the utter sterility of one of the forms of certain trimorphic species of Oxalis, when growing isolated. He has seen in St. Catharina, in Brazil, a large field of young sugar-cane, many acres in extent, covered with the red blossoms of one form alone, and these did not produce a single seed. His own land is covered with the short-styled form of a white-flowered trimorphic species, and this is equally sterile ; but when the three forms were planted near together in his garden they seeded freely. With two other trimorphic species he finds that isolated plants are always sterile.

Fritz Müller formerly believed that a species of Oxalis, which is so abundant in St. Catharina that it borders the roads for miles, was dimorphic instead of trimorphic. Although the pistils and stamens vary greatly in length, as was evident in some specimens sent to me, yet the plants can be divided into two sets, according to the lengths of these organs. A large proportion of the anthers are of a white colour and quite destitute of pollen ; others which are pale yellow contain many bad with some good grains ; and others again which are bright yellow have apparently sound pollen ; but he has never succeeded in finding any fruit on this species. The stamens in some of the flowers are partially converted into petals. Fritz Müller after reading my description, hereafter to be given, of the illegitimate offspring of various hetero-styled species, suspects that these plants of Oxalis may be the variable and sterile offspring of a single form of some trimorphic species, perhaps accidentally introduced into the district, which has since been propagated asexually. It is probable that this kind of propagation would be much aided by there being no expenditure in the production of seed.

Oxalis (Biophytum) sensitiva.—This plant is ranked by many botanists as a distinct genus. Mr. Thwaites sent me a number of flowers preserved in spirits from Ceylon, and they are clearly trimorphic. The style of the long-styled form is clothed with many scattered hairs, both simple and glandular; such hairs are much fewer on the style of the mid-styled, and quite absent from that of the short-styled form; so that this plant resembles in this respect *O. Valdiviana* and *Regnelli.* Calling the length of the two lobes of the stigma of the long-styled form 100, that of the mid-styled is 141, and that of the short-styled 164. In all other cases, in which the stigma in this genus differs in size in the three forms, the difference is of a reversed nature, the stigma of the long-styled being the largest, and that of the short-styled the smallest. The diameter of the pollen-grains from the longest stamens being represented by 100, those from the mid-length stamens are 91, and those from the shortest stamens 84 in diameter. This plant is remarkable, as we shall see in the last chapter of this volume, by producing long-styled, mid-styled, and short-styled cleistogamic flowers.

Homostyled Species of Oxalis.—Although the majority of the species in the large genus Oxalis seem to be trimorphic, some are homostyled, that is, exist under a single form; for instance the common *O. acetosella,* and according to Hildebrand two other widely distributed European species, *O. stricta* and *corniculata.* Fritz Müller also informs me that a similarly constituted species is found in St. Catharina, and that it is quite fertile with its own pollen when insects are excluded. The stigmas of *O. stricta* and of another homo-styled species, viz. *O. tropæoloides*, commonly stand on a level with the upper anthers, and both these species are likewise quite fertile when insects are excluded.

With respect to *O. acetosella*, Hildebrand says that in all the many specimens examined by him the pistil exceeded the longer stamens in length. I procured 108 flowers from the same number of plants growing in three distant parts of England ; of these 86 had their stigmas projecting considerably above, whilst 22 had them nearly on a level with the upper anthers. In one lot of 17 flowers from the same wood, the stigmas in every flower projected fully as much above the upper anthers as these stood above the lower anthers. So that these plants might fairly be compared with the long-styled form of a heterostyled species; and I at first thought that *O. acetosella* was trimorphic. But the case is one merely of great variability. The pollen-grains from the two sets of anthers, as observed by Hildebrand and myself, do not differ in diameter. I fertilised twelve flowers on several plants with pollen from a distinct plant, choosing those with pistils of a different length ; and 10 of these (i.e. 83 per cent.) produced capsules, which contained on an average 7·9 seeds. Fourteen flowers were fertilised with their own pollen, and 11 of these (i.e. 79 per cent.) yielded capsules, containing a larger average of seed, namely 9·2. These plants, therefore, in function show not the least sign of being heterostyled. I may add that 18 flowers protected by a net were left to fertilise themselves, and only 10 of these (i.e. 55 per cent.) yielded capsules, which contained on an average only 6·3 seeds. So that the access of insects, or artificial aid in placing pollen on the stigma, increases the fertility of the flowers; and I found that this applied especially to those having shorter pistils. It should be remembered that the flowers hang downwards, so that those with short pistils would be the least likely to receive their own pollen, unless they were aided in some manner.

Finally, as Hildebrand has remarked, there is no evidence that any of the heterostyled species of Oxalis are tending towards a diœcious condition, as Zuccarini and Lindley inferred from the differences in the reproductive organs of the three forms, the meaning of which they did not understand.

PONTEDERIA [SP. ?] (PONTEDERIACEÆ.)

Fritz Müller found this aquatic plant, which is allied to the Liliaceæ, growing in the greatest profusion on the banks of a river in Southern Brazil.* But only two forms were found, the flowers of which include three long and three short stamens. The pistil of the long-styled form, in two dried flowers which were sent me, was in length as 100 to 32, and its stigma as 100 to 80, compared with the same organs in the short-styled form. The long-styled stigma projects considerably above the upper anthers of the same flower, and stands on a level with the upper ones of the short-styled form. In the latter the stigma is seated beneath both its own sets of anthers, and is on a level with the anthers of the shorter stamens in the long-styled form. The anthers of the longer stamens of the short-styled form are to those of the shorter stamens of the long-styled form as 100 to 88 in length. The pollen-grains distended

---

* "Ueber den Trimorphismus   Zeitschrift,' &c., Band 6, 1871,
der   Pontederien"; 'Jenaische   p. 74.

with water from the longer stamens of the short-styled form are to those from the shorter stamens of the same form as 100 to 87 in diameter, as deduced from ten measurements of each kind. We thus see that the organs in these two forms differ from one another and are arranged in an analogous manner, as in the long and short-styled forms of the trimorphic species of Lythrum and Oxalis. Moreover, the longer stamens of the long-styled form of Pontederia, and the shorter ones of the short-styled form are placed in a proper position for fertilising the stigma of a mid-styled form. But Fritz Müller, although he examined a vast number of plants, could never find one belonging to the mid-styled form. The older flowers of the long-styled and short-styled plants had set plenty of apparently good fruit; and this might have been expected, as they could legitimately fertilise one another. Although he could not find the mid-styled form of this species, he possessed plants of another species growing in his garden, and all these were mid-styled; and in this case the pollen-grains from the anthers of the longer stamens were to those from the shorter stamens of the same flower as 100 to 86 in diameter, as deduced from ten measurements of each kind. These mid-styled plants growing by themselves never produced a single fruit.

Considering these several facts, there can hardly be a doubt that both these species of Pontederia are heterostyled and trimorphic. This case is an interesting one, for no other Monocotyledonous plant is known to be heterostyled. Moreover, the flowers are irregular, and all other heterostyled plants have almost symmetrical flowers. The two forms differ somewhat in the colour of their corollas, that of the short-styled being of a darker blue, whilst that of the long-styled tends towards violet, and no other such case is known. Lastly, the three longer stamens alternate with the three shorter ones, whereas in Lythrum and Oxalis the long and short stamens belong to distinct whorls. With respect to the absence of the mid-styled form in the case of the Pontederia which grows wild in Southern Brazil, this would probably follow if only two forms had been originally introduced there; for, as we shall hereafter see from the observations of Hildebrand, Fritz Müller and myself, when one form of Oxalis is fertilised exclusively by either of the other two forms, the offspring generally belong to the two parent-forms.

Fritz Müller has recently discovered, as he informs me, a third species of Pontederia, with all three forms growing together in pools in the interior of S. Brazil; so that no shadow of doubt can any longer remain about this genus including trimorphic species. He sent me dried flowers of all three forms. In the long-styled form the stigma stands a little above the tips of the petals, and on a level with the anthers of the longest stamens in the other two forms. The pistil is in length to that of the mid-styled as 100 to 56, and to that of the short-styled as 100 to 16. Its summit is rectangularly bent upwards, and the stigma is rather broader than that of the mid-styled, and broader in about the ratio of 7 to 4 than that of the short-styled. In the mid-styled form, the stigma is placed rather above the middle of the corolla, and nearly on a level with the mid-length stamens in the other two forms; its summit is a little bent upwards. In the short-styled form the pistil is, as we have seen, very short, and differs from that in the other two forms in being straight. It stands rather beneath the level of the anthers of the shortest stamens in the long-styled and mid-styled forms. The three anthers of each set of stamens, more especially those of the shortest stamens, are placed one beneath the other, and the ends of the filaments are bowed a little upwards, so that the pollen from all the anthers would be effectively brushed off by the proboscis of a visiting insect. The relative diameters of the pollen-grains, after having been long soaked in water, are given in the following list, as measured by my son Francis.

	Divisions of the Micrometer.
Long-styled form, from the mid-length stamens . . .	13·2
(Average of 20 measurements.)	
„ „ from the shortest stamens	9·0
(10 measurements.)	
Mid-styled form, from the longest stamens	16·4
(15 measurements.)	
„ „ from the shortest stamens	9·1
(20 measurements.)	
Short-styled form, from the longest stamens	14·6
(20 measurements.)	
„ „ from the mid-length stamens . . .	12·3
(20 measurements.)	

We have here the usual rule of the grains from the longer stamens, the tubes of which have to penetrate the longer pistil, being larger than those from the stamens of less length. The extreme difference in diameter between the grains from the longest stamens of the mid-styled form, and from the shortest stamens of the long-styled, is as 16·4 to 9·0, or as 100 to 55; and this is the greatest difference observed by me in any heterostyled plant. It is a singular fact that the grains from the corresponding longest stamens in the two forms differ considerably in diameter; as do those in a lesser degree from the corresponding mid-length stamens in the two forms; whilst those from the corresponding shortest stamens in the long- and mid-styled forms are almost exactly equal. Their inequality in the two first cases depends on the grains

in both sets of anthers in the short-styled form being
smaller than those from the corresponding anthers in
the other two forms ; and here we have a case parallel
with that of the mid-styled form of *Lythrum salicaria.*
In this latter plant the pollen-grains of the mid-styled
forms are of smaller size and have less fertilising power
than the corresponding ones in the other two forms ;
whilst the ovarium, however fertilised, yields a greater
number of seeds ; so that the mid-styled form is altogether more feminine in nature than the other two
forms.   In the case of Pontederia, the ovarium includes only a single ovule, and what the meaning of
the difference in size between the pollen-grains from
the corresponding sets of anthers may be, I will not
pretend to conjecture.

The clear evidence that the species just described is
heterostyled and trimorphic is the more valuable as
there is some doubt with respect to *P. cordata,* an inhabitant of the United States.   Mr. Leggett suspects[*]
that it is either dimorphic or trimorphic, for the
pollen-grains of the longer stamens are "more than
twice the diameter or than eight times the mass of
the grains of the shorter stamens. Though minute,
these smaller grains seem as perfect as the larger
ones."   On the other hand, he says that in all the
mature flowers, "the style was as long at least as
the longer stamens;" "whilst in the young flowers
it was intermediate in length between the two sets of
stamens;" and if this be so, the species can hardly be
heterostyled.

* 'Bull. of the Torrey Botanical Club,' 1875, vol. vi. p. 62.

CHAPTER V.

Illegitimate Offspring of Heterostyled Plants.

Illegitimate offspring from all three forms of Lythrum salicaria—Their dwarfed stature and sterility, some utterly barren, some fertile—Oxalis, transmission of form to the legitimate and illegitimate seedlings—Primula Sinensis, illegitimate offspring in some degree dwarfed and infertile—Equal-styled varieties of P. Sinensis, auricula, farinosa, and elatior—P. vulgaris, red-flowered variety, illegitimate seedlings sterile—P. veris, illegitimate plants raised during several successive generations, their dwarfed stature and sterility—Equal-styled varieties of P. veris—Transmission of form by Pulmonaria and Polygonum—Concluding remarks—Close parallelism between illegitimate fertilisation and hybridism.

We have hitherto treated of the fertility of the flowers of heterostyled plants, when legitimately and illegitimately fertilised. The present chapter will be devoted to the character of their offspring or seedlings. Those raised from legitimately fertilised seeds will be here called *legitimate seedlings* or *plants*, and those from illegitimately fertilised seeds, *illegitimate seedlings* or *plants*. They differ chiefly in their degree of fertility, and in their powers of growth or vigour. I will begin with trimorphic plants, and I must remind the reader that each of the three forms can be fertilised in six different ways; so that all three together can be fertilised in eighteen different ways. For instance, a long-styled form can be fertilised legitimately by the longest stamens of the mid-styled and short-styled forms, and illegitimately by its own-form mid-length and shortest stamens, also by the mid-length stamens of the mid-styled and by the shortest stamens of the short-styled form; so that the long-styled can be fertilised legitimately in two ways and illegitimately in four ways. The same holds good with respect to the mid-styled and short-styled forms. Therefore with trimorphic species six of the eighteen unions yield legitimate offspring, and twelve yield illegitimate offspring.

I will give the results of my experiments in detail, partly because the observations are extremely troublesome, and will not probably soon be repeated—thus, I was compelled to count under the microscope above 20,000 seeds of *Lythrum salicaria*—but chiefly because light is thus indirectly thrown on the important subject of hybridism.

LYTHRUM SALICARIA.

Of the twelve illegitimate unions two were completely barren, so that no seeds were obtained, and of course no seedlings could be raised. Seedlings were, however raised from seven of the ten remaining illegitimate unions. Such illegitimate seedlings when in flower were generally allowed to be freely and legitimately fertilised, through the agency of bees, by other illegitimate plants belonging to the two other forms growing close by. This is the fairest plan, and was usually followed; but in several cases (which will always be stated) illegitimate plants were fertilised with pollen taken from legitimate plants belonging to the other two forms; and this, as might have been expected, increased their fertility. *Lythrum salicaria* is much affected in its fertility by the nature of the season; and to avoid error from this source, as far as possible, my observations were continued during several years. Some few experiments were tried in 1863. The summer of 1864 was too hot and

dry, and, though the plants were copiously watered,
some few apparently suffered in their fertility, whilst
others were not in the least affected. The years
1865 and, especially, 1866, were highly favourable.
Only a few observations were made during 1867.
The results are arranged in classes according to the
parentage of the plants. In each case the average
number of seeds per capsule is given, generally taken
from ten capsules, which, according to my experience,
is a nearly sufficient number. The maximum number of seeds in any one capsule is also given; and
this is a useful point of comparison with the normal standard—that is, with the number of seeds
produced by legitimate plants legitimately fertilised. I will give likewise in each case the minimum
number. When the maximum and minimum differ
greatly, if no remark is made on the subject, it may
be understood that the extremes are so closely connected by intermediate figures that the average is a
fair one. Large capsules were always selected for
counting, in order to avoid over-estimating the infertility of the several illegitimate plants.

In order to judge of the degree of inferiority in
fertility of the several illegitimate plants, the following statement of the average and of the maximum
number of seeds produced by ordinary or legitimate
plants, when legitimately fertilised, some artificially
and some naturally, will serve as a standard of comparison, and may in each case be referred to. But I
give under each experiment the percentage of seeds
produced by the illegitimate plants, in comparison
with the standard legitimate number of the same
form. For instance, ten capsules from the illegitimate
long-styled plant (No. 10), which was legitimately
and naturally fertilised by other illegitimate plants,

contained on an average 44·2 seeds; whereas the capsules on legitimate long-styled plants, legitimately and naturally fertilised by other legitimate plants, contained on an average 93 seeds. Therefore this illegitimate plant yielded only 47 per cent. of the full and normal complement of seeds.

Standard Number of Seeds produced by Legitimate Plants of the three Forms, when legitimately fertilised.

Long-styled form : average number of seeds in each capsule, 93 ; maximum number observed out of twenty-three capsules, 159.

Mid-styled form: average number of seeds, 130; maximum number observed out of thirty-one capsules, 151.

Short-styled form: average number of seeds, 83·5 ; but we may, for the sake of brevity, say 83; maximum number observed out of twenty-five capsules, 112.

CLASSES I. and II. *Illegitimate Plants raised from Long-styled Parents fertilised with pollen from the mid-length or the shortest stamens of other plants of the same form.*

From this union I raised at different times three lots of illegitimate seedlings, amounting altogether to 56 plants. I must premise that, from not foreseeing the result, I did not keep a memorandum whether the eight plants of the first lot were the product of the mid-length or shortest stamens of the same form ; but I have good reason to believe that they were the product of the latter. These eight plants were much more dwarfed, and much more sterile than those in the other two lots. The latter were raised from a long-styled plant growing quite isolated, and fertilised by the agency of bees with its own pollen; and it is almost certain, from the relative position of the organs of fructification, that the stigma under these circumstances would receive pollen from the mid-length stamens.

All the fifty-six plants in these three lots proved long-styled; now, if the parent-plants had been legitimately fertilised by pollen from the longest stamens of the mid-styled and short-styled forms, only about one-third of the seedlings would have been long-styled, the other two-thirds being mid-styled and short-styled. In some other trimorphic and dimorphic genera we shall find the same curious fact, namely, that the long-styled form, fertilised illegitimately by its own-form pollen, produces almost exclusively long-styled seedlings.*

The eight plants of the first lot were of low stature: three which I measured attained, when fully grown, the heights of only 28, 29, and 47 inches; whilst legitimate plants growing close by were double this height, one being 77 inches. They all betrayed in their general appearance a weak constitution; they flowered rather later in the season, and at a later age than ordinary plants. Some did not flower every year; and one plant, behaving in an unprecedented manner, did not flower until three years old. In the two other lots none of the plants grew quite to their full and proper height, as could at once be seen by comparing them with the adjoining rows of legitimate plants. In several plants in all three lots, many of the anthers were either shrivelled or contained brown and tough, or pulpy

---

* Hildebrand first called attention ('Bot. Zeitung,' Jan. 1, 1864, p. 5) to this fact in the case of *Primula Sinensis*; but his results were not nearly so uniform as mine.

matter, without any good pollen-grains, and they never shed their contents ; they were in the state designated by Gärtner * as contabescent, which term I will for the future use. In one flower all the anthers were contabescent excepting two which appeared to the naked eye sound ; but under the microscope about two-thirds of the pollen-grains were seen to be small and shrivelled. In another plant, in which all the anthers appeared sound, many of the pollen-grains were shrivelled and of unequal sizes. I counted the seeds produced by seven plants (1 to 7) in the first lot of eight plants, probably the product of parents fertilised by their own-form shortest stamens, and the seeds produced by three plants in the other two lots, almost certainly the product of parents fertilised by their own-form mid-length stamens.

Plant 1. This long-styled plant was allowed during 1863 to be freely and legitimately fertilised by an adjoining illegitimate mid-styled plant, but it did not yield a single seed-capsule. It was then removed and planted in a remote place close to a brother long-styled plant No. 2, so that it must have been freely though illegitimately fertilised; under these circumstances it did not yield during 1864 and 1865 a single capsule. I should here state that a legitimate or ordinary long-styled plant, when growing isolated, and freely though illegitimately fertilised by insects with its own pollen, yielded an immense number of capsules, which contained on an average 21·5 seeds.

Plant 2. This long-styled plant, after flowering during 1863 close to an illegitimate mid-styled plant, produced less than twenty capsules, which contained on an average between four and five seeds. When subsequently growing in company with No. 1, by which it will have been illegitimately fertilised, it yielded in 1866 not a single capsule, but in 1865 it yielded twenty-two capsules: the best of these, fifteen in number, were examined; eight contained no seed, and the remaining seven contained on an average only three seeds, and these seeds were

---

* 'Beiträge zur Kenntniss der Befruchtung,' 1844, p. 116.

O

so small and shrivelled that I doubt whether they would have germinated.

Plants 3 and 4. These two long-styled plants, after being freely and legitimately fertilised during 1863 by the same illegitimate mid-styled plant as in the last case, were as miserably sterile as No. 2.

Plant 5. This long-styled plant, after flowering in 1863 close to an illegitimate mid-styled plant, yielded only four capsules, which altogether included only five seeds. During 1864, 1865, and 1866, it was surrounded either by illegitimate or legitimate plants of the other two forms; but it did not yield a single capsule. It was a superfluous experiment, but I likewise artificially fertilised in a legitimate manner twelve flowers; but not one of these produced a capsule; so that this plant was almost absolutely barren.

Plant 6. This long-styled plant, after flowering during the favourable year of 1866, surrounded by illegitimate plants of the other two forms, did not produce a single capsule.

Plant 7. This long-styled plant was the most fertile of the eight plants of the first lot. During 1865 it was surrounded by illegitimate plants of various parentage, many of which were highly fertile, and must thus have been legitimately fertilised. It produced a good many capsules, ten of which yielded an average of 36·1 seeds, with a maximum of 47 and a minimum of 22; so that this plant produced 39 per cent of the full number of seeds. During 1864 it was surrounded by legitimate and illegitimate plants of the other two forms; and nine capsules (one poor one being rejected) yielded an average of 41·9 seeds, with a maximum of 56 and a minimum of 28; so that, under these favourable circumstances, this plant, the most fertile of the first lot, did not yield, when legitimately fertilised, quite 45 per cent. of the full complement of seeds.

In the second lot of plants in the present class, descended from the long-styled form, almost certainly fertilised with pollen from its own mid-length stamens, the plants, as already stated, were not nearly so dwarfed or so sterile as in the first lot. All produced plenty of capsules. I counted the number of seeds in only three plants, viz. Nos. 8, 9, and 10.

Plant 8. This plant was allowed to be freely fertilised in 1864 by legitimate and illegitimate plants of the other two forms, and ten capsules yielded on an average 41·1 seeds, with a maximum of 73 and a minimum of 11. Hence this plant produced only 44 per cent. of the full complement of seeds.

Plant 9. This long-styled plant was allowed in 1865 to be freely fertilised by illegitimate plants of the other two forms, most of which were moderately fertile. Fifteen capsules yielded on an average 57·1 seeds, with a maximum of 86 and a minimum of 23. Hence the plant yielded 61 per cent. of the full complement of seeds.

Plant 10. This long-styled plant was freely fertilised at the same time and in the same manner as the last. Ten capsules yielded an average of 44·2 seeds, with a maximum of 69 and a minimum of 25; hence this plant yielded 47 per cent. of the full complement of seeds.

The nineteen long-styled plants of the third lot, of the same parentage as the last lot, were treated differently; for they flowered during 1867 by themselves so that they must have been illegitimately fertilised by one another. It has already been stated that a legitimate long-styled plant, growing by itself and visited by insects, yielded an average of 21·5 seeds per capsule, with a maximum of 35; but, to judge fairly of its fertility, it ought to have been observed during successive seasons. We may also infer from analogy that, if several legitimate long-styled plants were to fertilise one another, the average number of seeds would be increased; but how much increased I do not know; hence I have no perfectly fair standard of comparison by which to judge of the fertility of the three following plants of the present lot, the seeds of which I counted.

Plant 11. This long-styled plant produced a large crop of capsules, and in this respect was one of the most fertile of the whole lot of nineteen plants. But the average from ten capsules was only 35·9 seeds, with a maximum of 60 and a minimum of 8.

Plant 12. This long-styled plant produced very few capsules; and ten yielded an average of only 15·4 seeds, with a maximum of 30 and a minimum of 4.

Plant 13. This plant offers an anomalous case; it flowered profusely, yet produced very few capsules; but these contained numerous seeds. Ten capsules yielded an average of 71·9 seeds, with a maximum of 95 and a minimum of 29. Considering that this plant was illegitimate and illegitimately fertilised by its brother long-styled seedlings, the average and the maximum are so remarkably high that I cannot at all understand the case. We should remember that the average for a legitimate plant legitimately fertilised is 93 seeds.

CLASS III. *Illegitimate Plants raised from a Short-styled Parent fertilised with pollen from own-form mid-length stamen.*

I raised from this union nine plants, of which eight were short-styled and one long-styled; so that there seems to be a strong tendency in this form to reproduce, when self-fertilised, the parent-form; but the tendency is not so strong as with the long-styled. These nine plants never attained the full height of legitimate plants growing close to them. The anthers were contabescent in many of the flowers on several plants.

Plant 14. This short-styled plant was allowed during 1865 to be freely and legitimately fertilised by illegitimate plants descended from self-fertilised mid-, long- and short-styled plants. Fifteen capsules yielded an average of 28·3 seeds, with a maximum of 51 and a minimum of 11; hence this plant produced only 33 per cent. of the proper number of seeds. The seeds themselves were small and irregular in shape. Although so sterile on the female side, none of the anthers were contabescent.

Plant 15. This short-styled plant, treated like the last during the same year, yielded an average, from fifteen capsules, of 27 seeds, with a maximum of 49 and a minimum of 7. But two poor capsules may be rejected, and then the average rises to 32·6, with the same maximum of 49 and a minimum of 20; so that this plant attained 38 per cent. of the normal standard of fertility, and was rather more fertile than the last, yet many of the anthers were contabescent.

Plant 16. This short-styled plant, treated like the two last, yielded from ten capsules an average of 77·8 seeds, with a maximum of 97 and a minimum of 60; so that this plant produced 94 per cent. of the full number of seeds.

Plant 17. This, the one long-styled plant of the same parentage as the last three plants, when freely and legitimately fertilised in the same manner as the last, yielded an average from ten capsules of 76·3 rather poor seeds, with a maximum of 88 and a minimum of 57. Hence this plant produced 82 per cent. of the proper number of seeds. Twelve flowers enclosed in a net were artificially and legitimately fertilised with pollen from a legitimate short-styled plant; and nine capsules yielded an average of 82·5 seeds, with a maximum of 98 and a minimum of 51; so that its fertility was increased by the action of pollen from a legitimate plant, but still did not reach the normal standard.

CLASS IV. *Illegitimate Plants raised from a Mid-styled Parent fertilised with pollen from own-form longest stamens.*

After two trials, I succeeded in raising only four plants from this illegitimate union. These proved to be three mid-styled and one long-styled; but from so small a number we can hardly judge of the tendency in mid-styled plants when self-fertilised to reproduce the same form. These four plants never attained their full and normal height; the long-styled plant had several of its anthers contabescent.

Plant 18. This mid-styled plant, when freely and legitimately fertilised during 1865 by illegitimate plants descended from self-fertilised long-, short-, and mid-styled plants, yielded an average from ten capsules of 102·6 seeds, with a maximum of

131 and a minimum of 63: hence this plant did not produce quite 80 per cent. of the normal number of seeds. Twelve flowers were artificially and legitimately fertilised with pollen from a legitimate long-styled plant, and yielded from nine capsules an average of 116·1 seeds, which were finer than in the previous case, with a maximum of 135 and a minimum of 75; so that, as with Plant 17, pollen from a legitimate plant increased the fertility, but did not bring it up to the full standard.

Plant 19. This mid-styled plant, fertilised in the same manner and at the same period as the last, yielded an average from ten capsules of 73·4 seeds, with a maximum of 87 and a minimum of 64: hence this plant produced only 56 per cent. of the full number of seeds. Thirteen flowers were artificially and legitimately fertilised with pollen from a legitimate long-styled plant, and yielded ten capsules with an average of 95·6 seeds; so that the application of pollen from a legitimate plant added, as in the two previous cases, to the fertility, but did not bring it up to the proper standard.

Plant 20. This long-styled plant, of the same parentage with the two last mid-styled plants, and freely fertilised in the same manner, yielded an average from ten capsules of 69·6 seeds, with a maximum of 83 and a minimum of 52: hence this plant produced 75 per cent. of the full number of seeds.

Class V. *Illegitimate Plants raised from a Short-styled Parent fertilised with pollen from the mid-length stamens of the long-styled form.*

In the four previous classes, plants raised from the three forms fertilised with pollen from either the longer or shorter stamens of the same form, but generally not from the same plant, have been described. Six other illegitimate unions are possible, namely, between the three forms and the stamens in the other two forms which do not correspond in height with their pistils. But I succeeded in raising plants from only three of these six unions. From one of them, forming the present Class V., twelve plants were raised; these consisted of eight short-styled, and four long-styled plants, with not one mid-styled. These twelve plants never attained quite their full and proper height, but by no means deserved to be called dwarfs. The anthers in some of the flowers were contabescent. One plant was remarkable from all the longer stamens in every flower and from many of the shorter ones having their anthers in this condition. The pollen of four other plants, in which none of the anthers were contabescent, was examined; in one a moderate number of grains were minute and shrivelled, but in the other three they appeared perfectly sound. With respect to the power of producing seed, five plants (Nos. 21 to 25) were observed: one yielded scarcely more than half the normal number; a second was slightly infertile; but the three others actually produced a larger average number of seeds, with a higher maximum, than the standard. In my concluding remarks I shall recur to this fact, which at first appears inexplicable.

Plant 21. This short-styled plant, freely and legitimately fertilised during 1865 by illegitimate plants, descended from self-fertilised long-, mid- and short-styled parents, yielded an average from ten capsules of 43 seeds, with a maximum of 63 and a minimum of 26: hence this plant, which was the one with all its longer and many of its shorter stamens contabescent, produced only 52 per cent. of the proper number of seeds.

Plant 22. This short-styled plant produced perfectly sound pollen, as viewed under the microscope. During 1866 it was freely and legitimately fertilised by other illegitimate plants belonging to the present and the following class, both of which include many highly fertile plants. Under these circumstances it yielded from eight capsules an average of 100·5 seeds, with a maximum of 123 and a minimum of 86; so that it produced 121 per cent. of seeds in comparison with the normal standard. During 1864 it was allowed to be freely and legitimately fertilised by legitimate and illegitimate plants, and yielded an average, from eight capsules, of 104·2 seeds, with a maximum of 125 and a minimum of 90; consequently it exceeded the normal standard, producing 125 per cent. of seeds. In this case, as in some previous cases, pollen from legitimate plants added in a small degree to the fertility of the plant; and the fertility would, perhaps, have been still greater had not the summer of 1864 been very hot and certainly unfavourable to some of the plants of Lythrum.

Plant 23. This short-styled plant produced perfectly sound pollen. During 1866 it was freely and legitimately fertilised by the other illegitimate plants specified under the last experiment, and eight capsules yielded an average of 113·5 seeds, with a maximum of 123 and a minimum of 93. Hence this plant exceeded the normal standard, producing no less than 136 per cent. of seeds.

Plant 24. This long-styled plant produced pollen which seemed under the microscope sound; but some of the grains did not swell when placed in water. During 1864 it was legitimately fertilised by legitimate and illegitimate plants in the same manner as Plant 22, but yielded an average, from ten capsules, of only 55 seeds, with a maximum of 88 and a minimum of 24, thus attaining 59 per cent. of the normal fertility. This low degree of fertility, I presume, was owing to the unfavourable season; for during 1866, when legitimately fertilised by illegitimate plants in the manner described under No. 22, it yielded an average, from eight capsules, of 82 seeds, with a maximum of 120 and a minimum of 67, thus producing 88 per cent. of the normal number of seeds.

Plant 25. The pollen of this long-styled plant contained a moderate number of poor and shrivelled grains; and this is a surprising circumstance, as it yielded an extraordinary number of seeds. During 1866 it was freely and legitimately fertilised by illegitimate plants, as described under No. 22, and yielded an average, from eight capsules, of 122·5 seeds, with a maximum of 149 and a minimum of 84. Hence this plant exceeded the normal standard, producing no less than 131 per cent. of seeds.

Class VI. *Illegitimate Plants raised from Mid-styled Parents fertilised with pollen from the shortest stamens of the long-styled form.*

I raised from this union twenty-five plants, which proved to be seventeen long-styled and eight mid-

styled, but not one short-styled. None of these plants
were in the least dwarfed. I examined, during the
highly favourable season of 1866, the pollen of four
plants: in one mid-styled plant, some of the anthers of
the longest stamens were contabescent, but the pollen-
grains in the other anthers were mostly sound, as
they were in all the anthers of the shortest stamens;
in two other mid-styled and in one long-styled plant
many of the pollen-grains were small and shrivelled;
and in the latter plant as many as a fifth or sixth part
appeared to be in this state. I counted the seeds in
five plants (Nos. 26 to 30), of which two were moderately sterile and three fully fertile.

Plant 26. This mid-styled plant was freely and legitimately
fertilised, during the rather unfavourable year 1864, by numerous surrounding legitimate and illegitimate plants. It yielded
an average, from ten capsules, of 83·5 seeds, with a maximum
of 110 and a minimum of 64, thus attaining 64 per cent. of the
normal fertility. During the highly favourable year 1866, it
was freely and legitimately fertilised by illegitimate plants
belonging to the present Class and to Class V., and yielded
an average, from eight capsules, of 86 seeds, with a maximum
of 109 and a minimum of 61, and thus attained 66 per cent.
of the normal fertility. This was the plant with some of the
anthers of the longest stamens contabescent as above mentioned.

Plant 27. This mid-styled plant, fertilised during 1864 in the
same manner as the last, yielded an average, from ten capsules,
of 99·4 seeds, with a maximum of 122 and a minimum of 53,
thus attaining to 76 per cent. of the normal fertility. If the
season had been more favourable, its fertility would probably
have been somewhat greater, but, judging from the last experiment, only in a slight degree.

Plant 28. This mid-styled plant, when legitimately fertilised
during the favourable season of 1866, in the manner described
under No. 26, yielded an average, from eight capsules, of 89
seeds, with a maximum of 119 and a minimum of 69, thus producing 68 per cent. of the full number of seeds. In the pollen
of both sets of anthers, nearly as many grains were small and
shrivelled as sound.

Plant 29. This long-styled plant was legitimately fertilised during the unfavourable season of 1864, in the manner described under No. 26, and yielded an average, from ten capsules, of 84·6 seeds, with a maximum of 132 and a minimum of 47, thus attaining to 91 per cent. of the normal fertility. During the highly favourable season of 1866, when fertilised in the manner described under No. 26, it yielded an average, from nine capsules (one poor capsule having been excluded), of 100 seeds, with a maximum of 121 and a minimum of 77. This plant thus exceeded the normal standard, and produced 107 per cent. of seeds. In both sets of anthers there were a good many bad and shrivelled pollen-grains, but not so many as in the last-described plant.

Plant 30. This long-styled plant was legitimately fertilised during 1866 in the manner described under No. 26, and yielded an average, from eight capsules, of 94 seeds, with a maximum of 106 and a minimum of 66; so that it exceeded the normal standard, yielding 101 per cent. of seeds.

Plant 31. Some flowers on this long-styled plant were artificially and legitimately fertilised by one of its brother illegitimate mid-styled plants; and five capsules yielded an average of 90·6 seeds, with a maximum of 97 and a minimum of 79. Hence, as far as can be judged from so few capsules, this plant attained, under these favourable circumstances, 98 per cent. of the normal standard.

CLASS VII. *Illegitimate Plants raised from Mid-styled Parents fertilised with pollen from the longest stamens of the short-styled form.*

It was shown in the last chapter that the union from which these illegitimate plants were raised is far more fertile than any other illegitimate union; for the mid-styled parent, when thus fertilised, yielded an average (all very poor capsules being excluded) of 102·8 seeds, with a maximum of 130; and the seedlings in the present class likewise have their fertility not at all lessened. Forty plants were raised; and these attained their full height and were covered with seed-capsules.

Nor did I observe any contabescent anthers. It deserves, also, particular notice that these plants, differently from what occurred in any of the previous classes, consisted of all three forms, namely, eighteen short-styled, fourteen long-styled, and eight mid-styled plants. As these plants were so fertile, I counted the seeds only in the two following cases.

Plant 32. This mid-styled plant was freely and legitimately fertilised during the unfavourable year of 1864, by numerous surrounding legitimate and illegitimate plants. Eight capsules yielded an average of 127·2 seeds, with a maximum of 144 and a minimum of 96; so that this plant attained 98 per cent. of the normal standard.

Plant 33. This short-styled plant was fertilised in the same manner and at the same time with the last; and ten capsules yielded an average of 113·9, with a maximum of 137 and a minimum of 90. Hence this plant produced no less than 137 per cent. of seeds in comparison with the normal standard.

Concluding Remarks on the Illegitimate Offspring of the three forms of Lythrum salicaria.

From the three forms occurring in approximately equal numbers in a state of nature, and from the results of sowing seed naturally produced, there is reason to believe that each form, when legitimately fertilised, reproduces all three forms in about equal numbers. Now, we have seen (and the fact is a very singular one) that the fifty-six plants produced from the long-styled form, illegitimately fertilised with pollen from the same form (Class I. and II.), were all long-styled. The short-styled form, when self-fertilised (Class III.), produced eight short-styled and one long-styled plant; and the mid-styled form, similarly treated (Class IV.), produced three mid-styled and one long-styled offspring; so that these two forms, when illegitimately fertilised with pollen from the same form, evince a strong, but not exclusive, tendency to reproduce the parent-form. When the short-styled form was illegitimately fertilised by the long-styled form (Class V.), and again when the mid-styled was illegitimately fertilised by the long-styled (Class VI.), in each case the two parent-forms alone were reproduced. As thirty-seven plants were raised from these two unions, we may, with much confidence, believe that it is the rule that plants thus derived usually consist of both parent-forms, but not of the third form. When, however, the mid-styled form was illegitimately fertilised by the longest stamens of the short-styled (Class VII.), the same rule did not hold good; for the seedlings consisted of all three forms. The illegitimate union from which these latter seedlings were raised is, as previously stated, singularly fertile, and the seedlings themselves exhibited no signs of sterility and grew to their full height. From the consideration of these several facts, and from analogous ones to be given under Oxalis, it seems probable that in a state of nature the pistil of each form usually receives, through the agency of insects, pollen from the stamens of corresponding height from both the other forms. But the case last given shows that the application of two kinds of pollen is not indispensable for the production of all three forms. Hildebrand has suggested that the cause of all three forms being regularly and naturally reproduced, may be that some of the flowers are fertilised with one kind of pollen, and others on the same plant with the other kind of pollen. Finally, of the three forms, the long-styled evinces somewhat the strongest tendency to reappear amongst the offspring, whether both, or one, or neither of the parents are long-styled.

Table 30.

Tabulated results of the fertility of the foregoing illegitimate plants, when legitimately fertilised, generally by illegitimate plants, as described under each experiment. Plants 11, 12 and 13 are excluded, as they were illegitimately fertilised.

Normal Standard of Fertility of the three forms, when legitimately and naturally fertilised.

Form.	Average Number of Seeds per Capsule.	Maximum Number in any one Capsule.	Minimum Number in any one Capsule.
Long-styled	93	159	No record was kept, as all very poor capsules were rejected.
Mid-styled	130	151	
Short-styled . . .	83·5	112	

Class I. and II.—*Illegitimate Plants raised from Long-styled Parents fertilised with pollen from own-form mid-length or shortest stamens.*

Number of Plant.	Form.	Average Number of Seeds per Capsule.	Maximum Number in any one Capsule.	Minimum Number in any one Capsule.	Average Number of Seeds, expressed as the percentage of the Normal Standard.
Plant 1 . .	Long-styled	0	0	0	0
,, 2 . .	,,	4·5	?	0	5
,, 3 . .	,,	4·5	?	0	5
,, 4 . .	,,	4·5	?	0	5
,, 5 . .	,,	0 or 1	2	0	0 or 1
,, 6 . .	,,	0	0	0	0
,, 7 . .	,,	36·1	47	22	39
,, 8 . .	,,	41·1	73	11	44
,, 9 . .	,,	57·1	86	23	61
,, 10 . .	,,	44·2	69	25	47

Class III.—*Illegitimate Plants raised from Short-styled Parents fertilised with pollen from own-form shortest stamens.*

Plant 14 . .	Short-styled	28·3	51	11	33
,, 15 . .	,,	32·6	49	20	38
,, 16 . .	,,	77·8	97	60	94
,, 17 . .	Long-styled	76·3	88	57	82

TABLE 30—*continued.*

CLASS IV.—*Illegitimate Plants raised from Mid-styled Parents fertilised with pollen from own-form longest stamens.*

Number of Plant.	Form.	Average Number of Seeds per Capsule.	Maximum Number in any one Capsule.	Minimum Number in any one Capsule.	Average Number of Seeds, expressed as the percentage of the Normal Standard.
Plant 18 . .	Mid-styled.	102·6	131	63	80
„ 19 . .	„	73·4	87	64	56
„ 20 . .	Long-styled	69·6	83	52	75

CLASS V.—*Illegitimate Plants raised from Short-styled Parents fertilised with pollen from the mid-length stamens of the long-styled form.*

Plant 21 . .	Short-styled	43·0	63	26	52
„ 22 . .	„	100·5	123	86	121
„ 23 . .	„	113·5	123	93	136
„ 24 . .	Long-styled	82·0	120	67	88
„ 25 . .	„	122·5	149	84	131

CLASS VI.—*Illegitimate Plants raised from Mid-styled Parents fertilised with pollen from the shortest stamens of . the long-styled form.*

Plant 26 . .	Mid-styled.	86·0	109	61	66
„ 27 . .	„	99·4	122	53	76
„ 28 . .	„	89·0	119	69	68
„ 29 . .	Long-styled	100·0	121	77	107
„ 30 . .	„	94·0	106	66	101
„ 31 . .	„	90·6	97	79	98

CLASS VII.—*Illegitimate Plants raised from Mid-styled Parents fertilised with pollen from the longest stamens of the short-styled form.*

Plant 32 . .	Mid-styled.	127·2	144	96	98
„ 33 . .	Short-styled	113·9	137	90	137

The lessened fertility of most of these illegitimate plants is in many respects a highly remarkable phenomenon.   Thirty-three plants in the seven classes were subjected to various trials, and the seeds carefully counted.   Some of them were artificially fertilised, but the far greater number were freely fertilised (and this is the better and natural plan) through the agency of insects, by other illegitimate plants.   In the right-hand, or percentage column, in the preceding table, a wide difference in fertility between the plants in the first four and the last three classes may be perceived.   In the first four classes the plants are descended from the three forms illegitimately fertilised with pollen taken from the same form, but only rarely from the same plant.   It is necessary to observe this latter circumstance; for, as I have elsewhere shown,* most plants, when fertilised with their own pollen, or that from the same plant, are in some degree sterile, and the seedlings raised from such unions are likewise in some degree sterile, dwarfed, and feeble.   None of the nineteen illegitimate plants in the first four classes were completely fertile; one, however, was nearly so, yielding 96 per cent. of the proper number of seeds.   From this high degree of fertility we have many descending gradations, till we reach an absolute zero, when the plants, though bearing many flowers, did not produce, during successive years, a single seed or even seed-capsule.   Some of the most sterile plants did not even yield a single seed when legitimately fertilised with pollen from legitimate plants.   There is good reason to believe that the first seven plants in Class I. and II. were the offspring

---

* 'The Effects of Cross and Self-fertilisation in the Vegetable Kingdom,' 1876.

of a long-styled plant fertilised with pollen from its
own-form shortest stamens, and these plants were the
most sterile of all. The remaining plants in Class I.
and II. were almost certainly the product of pollen
from the mid-length stamens, and although very sterile, they were less so than the first set. None of the
plants in the first four classes attained their full and
proper stature; the first seven, which were the most
sterile of all (as already stated), were by far the most
dwarfed, several of them never reaching to half their
proper height. These same plants did not flower at so
early an age, or at so early a period in the season, as
they ought to have done. The anthers in many of
their flowers, and in the flowers of some other plants
in the first six classes, were either contabescent or included numerous small and shrivelled pollen-grains.
As the suspicion at one time occurred to me that the
lessened fertility of the illegitimate plants might be
due to the pollen alone having been affected, I may
remark that this certainly was not the case; for several
of them, when fertilised by sound pollen from legitimate plants, did not yield the full complement of
seeds; hence it is certain that both the female and
male reproductive organs were affected. In each of
the seven classes, the plants, though descended from
the same parents, sown at the same time and in the
same soil, differed much in their average degree of
fertility.

Turning now to the fifth, sixth, and seventh classes,
and looking to the right-hand column of the table, we
find nearly as many plants with a percentage of seeds
above the normal standard as beneath it. As with
most plants the number of seeds produced varies much,
it might be thought that the present case was one
merely of variability. But this view must be rejected,

as far as the less fertile plants in these three classes are concerned : first, because none of the plants in Class V. attained their proper height, which shows that they were in some manner affected ; and, secondly, because many of the plants in Classes V. and VI. produced anthers which were either contabescent or included small and shrivelled pollen-grains. And as in these cases the male organs were manifestly deteriorated, it is by far the most probable conclusion that the female organs were in some cases likewise affected, and that this was the cause of the reduced number of seeds.

With respect to the six plants in these three classes which yielded a very high percentage of seeds, the thought naturally arises that the normal standard of fertility for the long-styled and short-styled forms (with which alone we are here concerned) may have been fixed too low, and that the six illegitimate plants are merely fully fertile. The standard for the long-styled form was deduced by counting the seeds in twenty-three capsules, and for the short-styled form from twenty-five capsules. I do not pretend that this is a sufficient number of capsules for absolute accuracy ; but my experience has led me to believe that a very fair result may thus be gained. As, however, the maximum number observed in the twenty-five capsules of the short-styled form was low, the standard in this case may possibly be not quite high enough. But it should be observed, in the case of the illegitimate plants, that in order to avoid over-estimating their infertility, ten very fine capsules were always selected ; and the years 1865 and 1866, during which the plants in the three latter classes were experimented on, were highly favourable for seed-production. Now, if this

P

plan of selecting very fine capsules during favourable seasons had been followed for obtaining the normal standards, instead of taking, during various seasons, the first capsules which came to hand, the standards would undoubtedly have been considerably higher; and thus the fact of the six foregoing plants appearing to yield an unnaturally high percentage of seeds may, perhaps, be explained. On this view, these plants are, in fact, merely fully fertile, and not fertile to an abnormal degree. Nevertheless, as characters of all kinds are liable to variation, especially with organisms unnaturally treated, and as in the four first and more sterile classes, the plants derived from the same parents and treated in the same manner, certainly did vary much in sterility, it is possible that certain plants in the latter and more fertile classes may have varied so as to have acquired an abnormal degree of fertility. But it should be noticed that, if my standards err in being too low, the sterility of all the many sterile plants in the several classes will have to be estimated by so much the higher. Finally, we see that the illegitimate plants in the four first classes are all more or less sterile, some being absolutely barren, with one alone almost completely fertile; in the three latter classes, some of the plants are moderately sterile, whilst others are fully fertile, or possibly fertile in excess.

The last point which need here be noticed is that, as far as the means of comparison serve, some degree of relationship generally exists between the infertility of the illegitimate union of the several parent-forms and that of their illegitimate offspring. Thus the two illegitimate unions, from which the plants in Classes VI. and VII. were derived, yielded a fair amount of seed, and only a few of these plants are in any degree sterile. On the other hand, the illegitimate unions between plants of the same form always yield very few seeds, and their seedlings are very sterile. Long - styled parent-plants when fertilised with pollen from their own-form shortest stamens, appear to be rather more sterile than when fertilised with their own-form mid-length stamens; and the seedlings from the former union were much more sterile than those from the latter union. In opposition to this relationship, short-styled plants illegitimately fertilised with pollen from the mid-length stamens of the long-styled form (Class V.) are very sterile; whereas some of the offspring raised from this union were far from being highly sterile. It may be added that there is a tolerably close parallelism in all the classes between the degree of sterility of the plants and their dwarfed stature. As previously stated, an illegitimate plant fertilised with pollen from a legitimate plant has its fertility slightly increased. The importance of the several foregoing conclusions will be apparent at the close of this chapter, when the illegitimate unions between the forms of the same species and their illegitimate offspring, are compared with the hybrid unions of distinct species and their hybrid offspring.

OXALIS.

No one has compared the legitimate and illegitimate offspring of any trimorphic species in this genus. Hildebrand sowed illegitimately fertilised seeds of *Oxalis Valdiviana*,* but they did not germinate; and this fact, as he remarks, supports my view that an illegitimate union resembles a hybrid one between

---

* 'Bot Zeitung,' 1871, p. 433, footnote.

two distinct species, for the seeds in this latter case are often incapable of germination.

The following observations relate to the nature of the forms which appear among the legitimate seedlings of *Oxalis Valdiviana*. Hildebrand raised, as described in the paper just referred to, 211 seedlings from all six legitimate unions, and the three forms appeared among the offspring from each union. For instance, long-styled plants were legitimately fertilised with pollen from the longest stamens of the mid-styled form, and the seedlings consisted of 15 long-styled, 18 mid-styled, and 6 short-styled. We here see that a few short-styled plants were produced, though neither parent was short-styled; and so it was with the other legitimate unions. Out of the above 211 seedlings, 173 belonged to the same two forms as their parents, and only 38 belonged to the third form distinct from either parent. In the case of *O. Regnelli*, the result, as observed by Hildebrand, was nearly the same, but more striking: all the offspring from four of the legitimate unions consisted of the two parent-forms, whilst amongst the seedlings from the other two legitimate unions the third form appeared. Thus, of the 43 seedlings from the six legitimate unions, 35 belonged to the same two forms as their parents, and only 8 to the third form. Fritz Müller also raised in Brazil seedlings from long-styled plants of *O. Regnelli* legitimately fertilised with pollen from the longest stamens of the mid-styled form, and all these belonged to the two parent-forms.* Lastly, seedlings were raised by me from long-styled plants of *O. speciosa* legitimately fertilised by the short-styled form, and from the latter reciprocally fertilised by the long-styled; and these consisted of 33 long-styled and 26 short-styled plants, with not one mid-styled form. There can, therefore, be no doubt that the legitimate offspring from any two forms of Oxalis tend to belong to the same two forms as their parents; but that a few seedlings belonging to the third form occasionally make their appearance; and this latter fact, as Hildebrand remarks, may be attributed to atavism, as some of their progenitors will almost certainly have belonged to the third form.

When, however, any one form of Oxalis is fertilised illegiti-

---

* 'Jenaische Zeitschrift,' &c. Band vi. 1871, p. 75.

mately with pollen from the same form, the seedlings appear to belong invariably to this form. Thus Hildebrand states* that long-styled plants of *O. rosea* growing by themselves have been propagated in Germany year after year by seed, and have always produced long-styled plants. Again, 17 seedlings were raised from mid-styled plants of *O. hedysaroides* growing by themselves, and these were all mid-styled. So that the forms of Oxalis, when illegitimately fertilised wth their own pollen, behave like the long-styled form of *Lythrum salicaria*, which when thus fertilised always produced with me long-styled offspring.

PRIMULA.

PRIMULA SINENSIS.

I raised during February 1862, from some long-styled plants illegitimately fertilised with pollen from the same form, twenty-seven seedlings. These were all long-styled. They proved fully fertile or even fertile in excess; for ten flowers, fertilised with pollen from other plants of the same lot, yielded nine capsules, containing on an average 39·75 seeds, with a maximum in one capsule of 66 seeds. Four other flowers legitimately crossed with pollen from a legitimate plant, and four flowers on the latter crossed with pollen from the illegitimate seedlings, yielded seven capsules with an average of 53 seeds, with a maximum of 72. I must here state that I have found some difficulty in estimating the normal standard of fertility for the several unions of this species, as the results differ much during successive years, and the seeds vary so greatly in size that it is hard to

---

* 'Ueber den Trimorphismus in der Gattung Oxalis: Monatsberichte der Akad. der Wissen. zu Berlin,' 21st June 1866, p. 373; and 'Bot. Zeitung,' 1871, p. 435.

decide which ought to be considered good. In order
to avoid over-estimating the infertility of the several
illegitimate unions, I have taken the normal standard
as low as possible.

From the foregoing twenty-seven illegitimate plants,
fertilised with their own-form pollen, twenty-five seedling grandchildren were raised; and these were all
long-styled; so that from the two illegitimate generations fifty-two plants were raised, and all without
exception proved long-styled. These grandchildren
grew vigorously, and soon exceeded in height two
other lots of illegitimate seedlings of different parentage and one lot of equal-styled seedlings presently to
be described. Hence I expected that they would have
turned out highly ornamental plants; but when they
flowered, they seemed, as my gardener remarked, to
have gone back to the wild state; for the petals were
pale-coloured, narrow, sometimes not touching each
other, flat, generally deeply notched in the middle,
but not flexuous on the margin, and with the yellow
eye or centre conspicuous. Altogether these flowers
were strikingly different from those of their progenitors; and this, I think, can only be accounted
for on the principle of reversion. Most of the anthers
on one plant were contabescent. Seventeen flowers
on the grandchildren were illegitimately fertilised
with pollen taken from other seedlings of the same
lot, and produced fourteen capsules, containing on an
average 29·2 seeds; but they ought to have contained about 35 seeds. Fifteen flowers legitimately
fertilised with pollen from an illegitimate short-styled
plant (belonging to the lot next to be described) produced fourteen capsules, containing an average of 46
seeds; they ought to have contained at least 50 seeds.
Hence these grandchildren of illegitimate descent ap-

pear to have lost, though only in a very slight degree, their full fertility.

We will now turn to the short-styled form: from a plant of this kind, fertilised with its own-form pollen, I raised, during February 1862, eight seedlings, seven of which were short-styled and one long-styled. They grew slowly, and never attained to the full stature of ordinary plants; some of them flowered precociously, and others late in the season. Four flowers on these short-styled seedlings and four on the one long-styled seedling were illegitimately fertilised with their own-form pollen and produced only three capsules, containing on an average 23·6 seeds, with a maximum of 29; but we cannot judge of their fertility from so few capsules; and I have greater doubts about the normal standard for this union than about any other; but I believe that rather above 25 seeds would be a fair estimate. Eight flowers on these same short-styled plants, and the one long-styled illegitimate plant were reciprocally and legitimately crossed; they produced five capsules, which contained an average of 28·6 seeds, with a maximum of 36. A reciprocal cross between legitimate plants of the two forms would have yielded an average of at least 57 seeds, with a possible maximum of 74 seeds; so that these illegitimate plants were sterile when legitimately crossed.

I succeeded in raising from the above seven short-styled illegitimate plants, fertilised with their own-form pollen, only six plants—grandchildren of the first union. These, like their parents, were of low stature, and had so poor a constitution that four died before flowering. With ordinary plants it has been a rare event with me to have more than a single plant die out of a large lot. The two grandchildren which

lived and flowered were short-styled; and twelve of
their flowers were fertilised with their own-form pollen
and produced twelve capsules containing an average
of 28·2 seeds; so that these two plants, though belonging to so weakly a set, were rather more fertile
than their parents, and perhaps not in any degree
sterile. Four flowers on the same two grandchildren
were legitimately fertilised by a long-styled illegitimate plant, and produced four capsules, containing
only 32·2 seeds instead of about 64 seeds, which is
the normal average for legitimate short-styled plants
legitimately crossed.

By looking back, it will be seen that I raised at
first from a short-styled plant fertilised with its own-
form pollen one long-styled and seven short-styled
illegitimate seedlings. These seedlings were legitimately intercrossed, and from their seed fifteen plants
were raised, grandchildren of the first illegitimate
union, and to my surprise all proved short-styled.
Twelve short-styled flowers borne by these grandchildren were illegitimately fertilised with pollen
taken from other plants of the same lot, and produced
eight capsules which contained an average of 21·8
seeds, with a maximum of 35. These figures are
rather below the normal standard for such a union.
Six flowers were also legitimately fertilised with pollen
from an illegitimate long-styled plant and produced
only three capsules, containing on an average 23·6
seeds, with a maximum of 35. Such a union in the
case of a legitimate plant ought to have yielded an
average of 64 seeds, with a possible maximum of 73
seeds.

*Summary on the Transmission of Form, Constitution,
and Fertility of the Illegitimate Offspring of Primula
Sinensis.*—In regard to the long-styled plants, their

illegitimate offspring, of which fifty-two were raised in
the course of two generations, were all long-styled.*
These plants grew vigorously ; but the flowers in one
instance were small, appearing as if they had reverted
to the wild state. In the first illegitimate generation
they were perfectly fertile, and in the second their
fertility was only very slightly impaired. With
respect to the short-styled plants, twenty-four out of
twenty-five of their illegitimate offspring were short-
styled. They were dwarfed in stature, and one lot of
grandchildren had so poor a constitution that four out
of six plants perished before flowering. The two survivors, when illegitimately fertilised with their own-
form pollen, were rather less fertile than they ought
to have been; but their loss of fertility was clearly
shown in a special and unexpected manner, namely,
when legitimately fertilised by other illegitimate
plants : thus altogether eighteen flowers were fertilised
in this manner, and yielded twelve capsules, which
included on an average only 28·5 seeds, with a
maximum of 45. Now a legitimate short-styled plant
would have yielded, when legitimately fertilised, an
average of 64 seeds, with a possible maximum of 74.
This particular kind of infertility will perhaps be best
appreciated by a simile : we may assume that with
mankind six children would be born on an average from
an ordinary marriage ; but that only three would be
born from an incestuous marriage. According to the
analogy of *Primula Sinensis*, the children of such

---

* Dr. Hildebrand, who first
called attention to this subject
(' Bot. Zeitung,' 1864, p. 5), raised
from a similar illegitimate union
seventeen plants, of which fourteen were long-styled and three
short-styled. From a short-styled
plant illegitimately fertilised with
its own pollen he raised fourteen
plants, of which eleven were short-
styled and three long-styled.

incestuous marriages, if they continued to marry incestuously, would have their sterility only slightly increased; but their fertility would not be restored by a proper marriage; for if two children, both of incestuous origin, but in no degree related to each other, were to marry, the marriage would of course be strictly legitimate, nevertheless they would not give birth to more than half the full and proper number of children.

Equal-styled variety of Primula Sinensis.—As any variation in the structure of the reproductive organs, combined with changed function, is a rare event, the following cases are worth giving in detail. My attention was first called to the subject by observing, in 1862, a long-styled plant, descended from a self-fertilised long-styled parent, which had some of its flowers in an anomalous state, namely, with the stamens placed low down in the corolla as in the ordinary long-styled form, but with the pistils so short that the stigmas stood on a level with the anthers. These stigmas were nearly as globular and as smooth as in the short-styled form, instead of being elongated and rough as in the long-styled form. Here, then, we have combined in the same flower, the short stamens of the long-styled form with a pistil closely resembling that of the short-styled form. But the structure varied much even on the same umbel: for in two flowers the pistil was intermediate in length between that of the long and that of the short-styled form, with the stigma elongated as in the former, and smooth as in the latter; and in three other flowers the structure was in all respects like that of the long-styled form. These modifications appeared to me so remarkable that I fertilised eight of the flowers with their own pollen, and obtained five capsules, which contained on an average 43 seeds; and this number shows that the flowers had become abnormally fertile in comparison with those of ordinary long-styled plants when self-fertilised. I was thus led to examine the plants in several small collections, and the result showed that the equal-styled variety was not rare.

In a state of nature the long and short-styled forms would no doubt occur in nearly equal numbers, as I infer from the analogy of the other heterostyled species of Primula, and from having

TABLE 31.

Primula Sinensis.

Name of Owner or Place.	Long styled Form.	Short-styled Form.	Equal-styled Variety.
Mr. Horwood	0	0	17
Mr. Duck	20	0	9
Baston	30	18	15
Chichester	12	9	2
Holwood	42	12	0
High Elms	16	0	0
Westerham	1	5	0
My own plants from purchased seeds	13	7	0
Total	134	51	43

raised the two forms of the present species in exactly the same number from flowers which had been *legitimately* crossed.   The preponderance in the above table of the long-styled form over the short-styled (in the proportion of 134 to 51) results from gardeners generally collecting seed from self-fertilised flowers; and the long-styled flowers produce spontaneously much more seed (as shown in the first chapter) than the short-styled, owing to the anthers of the long-styled form being placed low down in the corolla, so that, when the flowers fall off, the anthers are dragged over the stigma; and we now also know that long-styled plants, when self-fertilised, very generally reproduce long-styled offspring.   From the consideration of this table, it occurred to me in the year 1862, that almost all the plants of the Chinese primrose cultivated in England would sooner or later become long-styled or equal-styled; and now, at the close of 1876, I have had five small collections of plants examined, and almost all consisted of long-styled, with some more or less well-characterised equal-styled plants, but with not one short-styled.

With respect to the equal-styled plants in the table, Mr. Horwood raised from purchased seeds four plants, which he remembered were certainly not long-styled, but either short or equal-styled, probably the latter.   These four plants were kept separate and allowed to fertilise themselves; from their seed the seventeen plants in the table were raised, all of which proved equal-styled.   The stamens stood low down in the corolla as in the long-styled form; and the stigmas, which were globular and

smooth, were either completely surrounded by the anthers, or
stood close above them. My son William made drawings for
me, by the aid of the camera, of the pollen of one of the above
equal-styled plants; and, in accordance with the position of the
stamens, the grains resembled in their small size those of the
long-styled form. He also examined pollen from two equal-styled
plants at Southampton; and in both of them the grains differed extremely in size in the same anthers, a large number
being small and shrivelled, whilst many were fully as large as
those of the short-styled form and rather more globular. It is
probable that the large size of these grains was due, not to their
having assumed the character of the short-styled form, but to
monstrosity; for Max Wichura has observed pollen-grains of
monstrous size in certain hybrids. The vast number of the
small shrivelled grains in the above two cases explains the fact
that, though equal-styled plants are generally fertile in a high
degree, yet some of them yield few seeds. I may add that my
son compared, in 1875, the grains from two white-flowered
plants, in both of which the pistil projected above the anthers,
but neither were properly long-styled or equal-styled; and in
the one in which the stigma projected most, the grains were
in diameter to those in the other plant, in which the stigma projected less, as 100 to 88; whereas the difference between the
grains from perfectly characterised long-styled and short-styled
plants is as 100 to 57. So that these two plants were in an
intermediate condition. To return to the 17 plants in the first
line of Table 31: from the relative position of their stigmas and
anthers, they could hardly fail to fertilise themselves; and accordingly four of them spontaneously yielded no less than 180
capsules; of these Mr. Horwood selected eight fine capsules for
sowing; and they included on an average 54·8 seeds, with a
maximum of 72. He gave me thirty other capsules, taken
by hazard, of which twenty-seven contained good seeds, aver-
aging 35·5, with a maximum of 70; but if six poor capsules, each with less than 13 seeds, be excluded, the average
rises to 42·5. These are higher numbers than could be expected from either well-characterised form if self-fertilised; and
this high degree of fertility accords with the view that the
male organs belonged to one form, and the female organs partially to the other form; so that a self-union in the case of the
equal-styled variety is in fact a legitimate union.

The seed saved from the above seventeen self-fertilised equal-styled plants produced sixteen plants, which all proved equal-styled, and resembled their parents in all the above-specified respects. The stamens, however, in one plant were seated higher up the tube of the corolla than in the true long-styled form; in another plant almost all the anthers were contabescent. These sixteen plants were the grandchildren of the four original plants, which it is believed were equal-styled; so that this abnormal condition was faithfully transmitted, probably through three, and certainly through two generations. The fertility of one of these grandchildren was carefully observed: six flowers were fertilised with pollen from the same flower, and produced six capsules, containing on an average 68 seeds, with a maximum of 82, and a minimum of 40. Thirteen capsules spontaneously self-fertilised yielded an average of 53·2 seeds, with the astonishing maximum in one of 97 seeds. In no legitimate union has so high an average as 68 seeds been observed by me, or nearly so high a maximum as 82 and 97. These plants, therefore, not only have lost their proper heterostyled structure and peculiar functional powers, but have acquired an abnormal grade of fertility—unless, indeed, their high fertility may be accounted for by the stigmas receiving pollen from the circumjacent anthers at exactly the most favourable period.

With respect to Mr. Duck's lot in Table 31, seed was saved from a single plant, of which the form was not observed, and this produced nine equal-styled and twenty long-styled plants. The equal-styled resembled in all respects those previously described; and eight of their capsules spontaneously self-fertilised contained on an average 44·4 seeds, with a maximum of 61 and a minimum of 23. In regard to the twenty long-styled plants, the pistil in some of the flowers did not project quite so high as in ordinary long-styled flowers; and the stigmas, though properly elongated, were smooth; so that we have here a slight approach in structure to the pistil of the short-styled form. Some of these long-styled plants also approached the equal-styled in function; for one of them produced no less than fifteen spontaneously self-fertilised capsules, and of these eight contained, on an average, 31·7 seeds, with a maximum of 61. This average would be rather low for a long-styled plant artificially fertilised with its own pollen, but is high for one spontaneously self-fertilised. For instance, thirty-four capsules produced by the illegitimate grandchildren of a long-styled plant, spontaneously self-fertilised, contained on an average only 9·1 seeds, with a maximum of 46. Some seeds indiscriminately saved from the foregoing twenty-nine equal-styled and long-styled plants produced sixteen seedlings, grandchildren of the original plant belonging to Mr. Duck; and these consisted of fourteen equal-styled and two long-styled plants; and I mention this fact as an additional instance of the transmission of the equal-styled variety.

The third lot in the table, namely the Baston plants, are the last which need be mentioned. The long and short-styled plants, and the fifteen equal-styled plants, were descended from two distinct stocks. The latter were derived from a single plant, which the gardener is positive was not long-styled; hence, probably, it was equal-styled. In all these fifteen plants the anthers, occupying the same position as in the long-styled form, closely surrounded the stigma, which in one instance alone was slightly elongated. Notwithstanding this position of the stigma, the flowers, as the gardener assured me, did not yield many seeds; and this difference from the foregoing cases may perhaps have been caused by the pollen being bad, as in some of the South-ampton equal-styled plants.

Conclusions with respect to the equal-styled variety of P. Sinensis.—That this is a variation, and not a third or distinct form, as in the trimorphic genera *Lythrum* and *Oxalis,* is clear; for we have seen its first appearance in one out of a lot of illegitimate long-styled plants; and in the case of Mr. Duck's seedlings, long-styled plants, only slightly deviating from the normal state, as well as equal-styled plants were produced from the same self-fertilised parent. The position of the sta-mens in their proper place low down in the tube of the corolla, together with the small size of the pollen-grains, show, firstly, that the equal-styled variety is a modification of the long-styled form, and, secondly, that the pistil is the part which has varied most, as indeed was obvious in many of the plants. This variation is of frequent occurrence, and is strongly inherited when ti has once appeared. It would, however, have pos-

sessed little interest if it had consisted of a mere
change of structure; but this is accompanied by modified fertility. Its occurrence apparently stands in
close relation with the illegitimate birth of the parent
plant; but to this whole subject I shall hereafter
recur.

PRIMULA AURICULA.

Although I made no experiments on the illegitimate offspring
of this species, I refer to it for two reasons:—First, because
I have observed two equal-styled plants in which the pistil
resembled in all respects that of the long-styled form, whilst
the stamens had become elongated as in the short-styled form,
so that the stigma was almost surrounded by the anthers. The
pollen-grains, however, of the elongated stamens resembled in
their small size those of the shorter stamens proper to the long-
styled form. Hence these plants have become equal-styled by
the increased length of the stamens, instead of, as with P.
Sinensis, by the diminished length of the pistil. Mr. J. Scott
observed five other plants in the same state, and he shows[*] that
one of them, when self-fertilised, yielded more seed than an
ordinary long- or short-styled form would have done when
similarly fertilised, but that it was far inferior in fertility to
either form when legitimately crossed. Hence it appears that
the male and female organs of this equal-styled variety have
been modified in some special manner, not only in structure,
but in functional powers. This, moreover, is shown by the
singular fact that both the long-styled and short-styled plants,
fertilised with pollen from the equal-styled variety, yield a
lower average of seed than when these two forms are fertilised
with their own pollen.

The second point which deserves notice is that florists always
throw away the long-styled plants, and save seed exclusively
from the short-styled form. Nevertheless, as Mr. Scott was
informed by a man who raises this species extensively in Scotland, about one-fourth of the seedlings appear long-styled; so
that the short-styled form of the Auricula, when fertilised by
its own pollen, does not reproduce the same form in so large a
proportion as in the case of P. Sinensis. We may further infer

---

[*] 'Journal Proc. Linn. Soc.' viii. (1864) p. 91.

that the short-styled form is not rendered quite sterile by a
long course of fertilisation with pollen of the same form : but as
there would always be some liability to an occasional cross with
the other form, we cannot tell how long self-fertilisation has
been continued.

PRIMULA FARINOSA.

Mr. Scott says * that it is not at all uncommon to find equal-
styled plants of this heterostyled species. Judging from the
size of the pollen-grains, these plants owe their structure, as in
the case of *P. auricula*, to the abnormal elongation of the
stamens of the long-styled form. In accordance with this view,
they yield less seed when crossed with the long-styled form
than with the short-styled. But they differ in an anomalous
manner from the equal-styled plants of *P. auricula* in being
extremely sterile with their own pollen.

PRIMULA ELATIOR.

It was shown in the first chapter, on the authority of
Herr Breitenbach, that equal-styled flowers are occasionally
found on this species whilst growing in a state of nature ; and
this is the only instance of such an occurrence known to me,
with the exception of some wild plants of the Oxlip—a hybrid
between *P. veris* and *vulgaris*—which were equal-styled. Herr
Breitenbach's case is remarkable in another way ; for equal-
styled flowers were found in two instances on plants which bore
both long-styled and short-styled flowers. In every other
instance these two forms and the equal-styled variety have been
produced by distinct plants.

PRIMULA VULGARIS, Brit. Fl.

Var. *acaulis* of Linn. and *P. acaulis* of Jacq.

Var. *rubra*.—Mr. Scott states† that this variety, which
grew in the Botanic Garden in Edinburgh, was quite
sterile when fertilised with pollen from the common
primrose, as well as from a white variety of the same

---

* 'Journal Proc. Linn. Soc.' viii. (1864), p. 115.
† Ibid. p. 98.

species, but that some of the plants, when artificially
fertilised with their own pollen, yielded a moderate
supply of seed. He was so kind as to send me some
of these self-fertilised seeds, from which I raised the
plants immediately to be described. I may premise
that the results of my experiments on the seedlings,
made on a large scale, do not accord with those by
Mr. Scott on the parent-plant.

First, in regard to the transmission of form and
colour. The parent-plant was long-styled, and of a
rich purple colour. From the self-fertilised seed 23
plants were raised; of these 18 were purple of different shades, with 2 of them a little streaked and
freckled with yellow, thus showing a tendency to
reversion; and 5 were yellow, but generally with a
brighter orange centre than in the wild flower. All
the plants were profuse flowerers. All were long-
styled; but the pistil varied a good deal in length
even on the same plant, being rather shorter, or considerably longer, than in the normal long-styled form;
and the stigmas likewise varied in shape. It is,
therefore, probable that an equal-styled variety of the
primrose might be found on careful search; and I
have received two accounts of plants apparently in this
condition. The stamens always occupied their proper
position low down in the corolla; and the pollen-
grains were of the small size proper to the long-styled
form, but were mingled with many minute and
shrivelled grains. The yellow-flowered and the purple-
flowered plants of this first generation were fertilised
under a net with their own pollen, and the seed separately sown. From the former, 22 plants were raised,
and all were yellow and long-styled. From the latter
or the purple-flowered plants, 24 long-styled plants
were raised, of which 17 were purple and 7 yellow.

Q

In this last case we have an instance of reversion in colour, without the possibility of any cross, to the grand-parents or more distant progenitors of the plants in question. Altogether 23 plants in the first generation and 46 in the second generation were raised; and the whole of these 69 illegitimate plants were long-styled!

Eight purple-flowered and two yellow-flowered plants of the first illegitimate generation were fertilised in various ways with their own pollen and with that of the common primrose; and the seeds were separately counted, but as I could detect no difference in fertility between the purple and yellow varieties, the results are run together in the following table. (See next page.)

If we compare the figures in this table with those given in the first chapter, showing the normal fertility of the common primrose, we shall see that the illegiti-mate purple- and yellow-flowered varieties are very sterile. For instance, 72 flowers were fertilised with their own pollen and produced only 11 good capsules; but by the standard they ought to have produced 48 capsules; and each of these ought to have contained on an average 52·2 seeds, instead of only 11·5 seeds. When these plants were illegitimately and legiti-mately fertilised with pollen from the common prim-rose, the average numbers were increased, but were far from attaining the normal standards. So it was when both forms of the common primrose were fertilised with pollen from these illegitimate plants; and this shows that their male as well as their female organs were in a deteriorated condition. The sterility of these plants was shown in another way, namely, by their not producing any capsules when the access of all insects (except such minute ones as Thrips) was prevented; for under these circumstances the common long-styled

TABLE 32.

Primula vulgaris.

Nature of Plant experimented on, and kind of Union.	Number of Flowers fertilised.	Number of Capsules produced.	Average Number of Seeds per Capsule.	Maximum Number of Seeds in any one Capsule.	Minimum Number of Seeds in any one Capsule.
Purple- and yellow-flowered illegitimate long-styled plants, *illegitimately* fertilised with pollen from the same plant . . .	72	11	11·5	26	5
Purple- and yellow-flowered illegitimate long-styled plants, *illegitimately* fertilised with pollen from the common long-styled primrose.	72	39	31·4	62	3
Or, if the ten poorest capsules, including less than 15 seeds, be rejected, we get.	72	29	40·6	62	18
Purple- and yellow-flowered illegitimate long-styled plants, *legitimately* fertilised with pollen from the common short-styled primrose.	26	18	36·4	60	9
Or, if the two poorest capsules, including less than 15 seeds, be rejected, we get.	26	16	41·2	60	15
The long-styled form of the common primrose, *illegitimately* fertilised with pollen from the long-styled illegitimate purple- and yellow-flowered plants	20	14	15·4	46	1
Or, if the three poorest capsules be rejected, we get	20	11	18·9	46	8
The short-styled form of the common primrose, *legitimately* fertilised with pollen from the long-styled illegitimate purple- and yellow-flowered plants	10	6	30·5	61	6

primrose produces a considerable number of capsules.
There can, therefore, be no doubt that the fertility of

these plants was greatly impaired.   The loss is not
correlated with the colour of the flower ; and it was to
ascertain this point that I made so many experiments.
As the parent-plant growing in Edinburgh was found
by Mr. Scott to be in a high degree sterile, it may
have transmitted a similar tendency to its offspring,
independently of their illegitimate birth.   I am, however, inclined to attribute some weight to the illegiti-
macy of their descent, both from the analogy of other
cases, and more especially from the fact that when the
plants were *legitimately* fertilised with pollen of the
common primrose they yielded an average, as may be
seen in the table, of only 5 more seeds than when
illegitimately fertilised with the same pollen.   Now we
know that it is eminently characteristic of the illegitimate offspring of *Primula Sinensis* that they yield but
few more seeds when legitimately fertilised than when
fertilised with their own-form pollen.

Primula veris, Brit. Fl.

Var. *officinalis* of Linn., *P. officinalis* of Jacq.

Seeds from the short-styled form of the cowslip
fertilised with pollen from the same form germinate
so badly that I raised from three successive sowings
only fourteen plants, which consisted of nine short-
styled and five long-styled plants.   Hence the short-
styled form of the cowslip, when self-fertilised, does not
transmit the same form nearly so truly as does that
of *P. Sinensis*.   From the long-styled form, always
fertilised with its own-form pollen, I raised in the
first generation three long-styled plants,—from their
seed 53 long-styled grandchildren,—from their seed
4 long-styled great-grandchildren,—from their seed
20 long-styled great-great-grandchildren,—and lastly,

from their seed 8 long-styled and 2 short-styled great-great-great-grandchildren. In this last generation short-styled plants appeared for the first time in the course of the six generations,—the parent long-styled plant which was fertilised with pollen from another plant of the same form being counted as the first generation. Their appearance may be attributed to atavism. From two other long-styled plants, fertilised with their own-form pollen, 72 plants were raised, which consisted of 68 long-styled and 4 short-styled. So that altogether 162 plants were raised from illegitimately fertilised long-styled cowslips, and these consisted of 156 long-styled and 6 short-styled plants.

We will now turn to the fertility and powers of growth possessed by the illegitimate plants. From a short-styled plant, fertilised with its own-form pollen, one short-styled and two long-styled plants, and from a long-styled plant similarly fertilised three long-styled plants were at first raised. The fertility of these six illegitimate plants was carefully observed; but I must premise that I cannot give any satisfactory standard of comparison as far as the number of the seeds is concerned; for though I counted the seeds of many legitimate plants fertilised legitimately and illegitimately, the number varied so greatly during successive seasons that no one standard will serve well for illegitimate unions made during different seasons. Moreover the seeds in the same capsule frequently differ so much in size that it is scarcely possible to decide which ought to be counted as good seed. There remains as the best standard of comparison the proportional number of fertilised flowers which produce capsules containing any seed.

First, for the one illegitimate short-styled plant. In the course of three seasons 27 flowers were illegitimately fertilised with pollen from the same plant, and they yielded only a single capsule, which, however, contained a rather large number of seeds for a union of this nature, namely, 23. As a standard of comparison I may state that during the same three seasons 44 flowers borne by legitimate short-styled plants were self-fertilised, and yielded 26 capsules; so that the fact of the 27 flowers on the illegitimate plant having produced only one capsule proves how sterile it was. To show that the conditions of life were favourable, I will add that numerous plants of this and other species of Primula all produced an abundance of capsules whilst growing close by in the same soil with the present and following plants. The sterility of the above illegitimate short-styled plant depended on both the male and female organs being in a deteriorated condition. This was manifestly the case with the pollen; for many of the anthers were shrivelled or contabescent. Nevertheless some of the anthers contained pollen, with which I succeeded in fertilising some flowers on the illegitimate long-styled plants immediately to be described. Four flowers on this same short-styled plant were likewise *legitimately* fertilised with pollen from one of the following long-styled plants; but only one capsule was produced, containing 26 seeds; and this is a very low number for a legitimate union.

With respect to the five illegitimate long-styled plants of the first generation, derived from the above self-fertilised short-styled and long-styled parents, their fertility was observed during the same three years. These five plants, when self-fertilised, differed considerably from one another in their degree of fertility, as was the case with the illegitimate long-styled plants of *Lythrum salicaria;* and their fertility varied much according to the season. I may premise, as a standard of comparison, that during the same years 56 flowers on legitimate long-styled plants of the same age and grown in the same soil, were fertilised with their own pollen, and yielded 27 capsules; that is, 48 per cent. On one of the five illegitimate long-styled plants 36 flowers were self-fertilised in the course of the three years, but they did not produce a single capsule. Many of the anthers on this plant were contabescent; but some seemed to contain sound pollen. Nor were the female organs quite impotent; for I obtained from a *legitimate* cross one capsule with good seed. On a second illegitimate long-styled plant 44 flowers were fertilised during the same years with their own pollen, but they produced only a single capsule. The third and fourth plants were in a very slight degree more productive. The fifth· and last plant was decidedly more fertile; for 42 self-fertilised flowers yielded 11 capsules. Altogether, in the course of the three years, no less than 160 flowers on these five illegitimate long-styled plants were fertilised with their own pollen, but they yielded only 22 capsules. According to the standard above given, they ought to have yielded 80 capsules. These 22 capsules contained on an average 15·1 seeds. I believe, subject to the doubts before specified, that with legitimate plants the average number from a union of this nature would have been above 20 seeds. Twenty-four flowers on these same five illegitimate long-styled plants were legitimately fertilised with pollen from the above-described illegitimate short-styled plant, and produced only 9 capsules, which is an extremely small number for a legitimate union. These 9 capsules, however, contained an average of 38 apparently good seeds, which is as large a number as legitimate plants sometimes yield. But this high average was almost certainly false; and I mention the case for the sake of showing the difficulty of arriving at a fair result; for this average mainly depended on two capsules containing the extraordinary numbers of 75 and 56 seeds; these seeds, however, though I felt bound to count them, were so poor that, judging from trials made in other cases, I do not suppose that one would have germinated; and therefore they ought not to have been included. Lastly, 20 flowers were legitimately fertilised with pollen from a legitimate plant, and this increased their fertility; for they produced 10 capsules. Yet this is but a very small proportion for a legitimate union.

There can, therefore, be no doubt that these five long-styled plants and the one short-styled plant of the first illegitimate generation were extremely sterile. Their sterility was shown, as in the case of hybrids, in another way, namely, by their flowering profusely, and especially by the long endurance of the flowers. For instance, I fertilised many flowers on these plants, and fifteen days afterwards (viz. on March 22nd) I fertilised numerous long-styled and short-styled flowers on common cowslips growing close by. These latter flowers, on April 8th, were withered, whilst most of the illegitimate flowers remained quite fresh for several days subsequently; so that some of these illegitimate plants, after being fertilised, remained in full bloom for above a month.

We will now turn to the fertility of the 53 illegitimate long-styled grandchildren, descended from the long-styled plant which was first fertilised with its own pollen. The pollen in two of these plants included a multitude of small and shrivelled grains. Nevertheless they were not very sterile; for 25 flowers, fertilised with their own pollen, produced 15 capsules, containing an average of 16·3 seeds. As already stated, the probable average with legitimate plants for a union of this nature is rather above 20 seeds. These plants were remarkably healthy and vigorous, as long as they were kept under highly favourable conditions in pots in the greenhouse; and such treatment greatly increases the fertility of the cowslip. When these same plants were planted during the next year (which, however, was an unfavourable one), out of doors in good soil, 20 self-fertilised flowers produced only 5 capsules, containing extremely few and wretched seeds.

Four long-styled great-grandchildren were raised from the self-fertilised grandchildren, and were kept under the same highly favourable conditions in the greenhouse; 10 of their flowers were fertilised with own-form pollen and yielded the large proportion of 6 capsules, containing on an average 18·7 seeds. From these seeds 20 long-styled great-great-grandchildren were raised, which were likewise kept in the greenhouse. Thirty of their flowers were fertilised with their own pollen and yielded 17 capsules, containing on an average no less than 32, mostly fine seeds. It appears, therefore, that the fertility of these plants of the fourth illegitimate generation, as long as they were kept under highly favourable conditions, had not decreased, but had rather increased. The result, however, was widely different when they were planted out of doors in good soil, where other cowslips grew vigorously and were completely fertile; for these illegitimate plants now became much dwarfed in stature and extremely sterile, notwithstanding that they were exposed to the visits of insects, and must have been legitimately fertilised by the surrounding legitimate plants. A whole

row of these plants of the fourth illegitimate generation, thus freely exposed and legitimately fertilised,
produced only 3 capsules, containing on an average
only 17 seeds. During the ensuing winter almost all
these plants died, and the few survivors were miserably
unhealthy, whilst the surrounding legitimate plants
were not in the least injured.

The seeds from the great-great-grandchildren were
sown, and 8 long-styled and 2 short-styled plants of
the fifth illegitimate generation raised. These whilst
still in the greenhouse produced smaller leaves and
shorter flower-stalks than some legitimate plants with
which they grew in competition; but it should be observed that the latter were the product of a cross with
a fresh stock,—a circumstance which by itself would
have added much to their vigour.* When these illegitimate plants were transferred to fairly good soil
out of doors, they became during the two following
years much more dwarfed in stature and produced very
few flower-stems; and although they must have been
legitimately fertilised by insects, they yielded capsules, compared with those produced by the surrounding legitimate plants, in the ratio only of 5 to 100!
It is therefore certain that illegitimate fertilisation,
continued during successive generations, affects the
powers of growth and fertility of *P. veris* to an extraordinary degree; more especially when the plants are
exposed to ordinary conditions of life, instead of being
protected in a greenhouse.

Equal-styled red variety of P. veris.—Mr. Scott has described †
a plant of this kind growing in the Botanic Garden of Edinburgh. He states that it was highly self-fertile, although insects

---

* For full details of this experiment, see my ' Effects of Cross
and Self-fertilisation,' 1876, p. 220.

† ' Proc. Linn. Soc.' vol. viii.
(1864), p. 105.

were excluded; and he explains this fact by showing, first, that
the anthers and stigma are in close apposition, and that the
stamens in length, position and size of their pollen-grains
resemble those of the short-styled form, whilst the pistil resembles that of the long-styled form both in length and in the
structure of the stigma. Hence the self-union of this variety is,
in fact, a legitimate union, and consequently is highly fertile.
Mr. Scott further states that this variety yielded very few seeds
when fertilised by either the long- or short-styled common
cowslip, and, again, that both forms of the latter, when fertilised
by the equal-styled variety, likewise produced very few seeds.
But his experiments with the cowslip were few, and my results
do not confirm his in any uniform manner.

I raised twenty plants from self-fertilised seed sent me by Mr.
Scott; and they all produced red flowers, varying slightly in
tint. Of these, two were strictly long-styled both in structure
and in function; for their reproductive powers were tested by
crosses with both forms of the common cowslip. Six plants
were equal-styled; but on the same plant the pistil varied a
good deal in length during different seasons. This was likewise
the case, according to Mr. Scott, with the parent-plant. Lastly,
twelve plants were in appearance short-styled; but they varied
much more in the length of their pistils than ordinary short-
styled cowslips, and they differed widely from the latter in
their powers of reproduction. Their pistils had become short-
styled in structure, whilst remaining long-styled in function.
Short-styled cowslips, when insects are excluded, are extremely
barren: for instance, on one occasion six fine plants produced
only about 50 seeds (that is, less than the product of two good
capsules), and on another occasion not a single capsule. Now,
when the above twelve apparently short-styled seedlings were
similarly treated, nearly all produced a great abundance of
capsules, containing numerous seeds, which germinated remarkably well. Moreover three of these plants, which during
the first year were furnished with quite short pistils, on the
following year produced pistils of extraordinary length. The
greater number, therefore, of these short-styled plants could not
be distinguished in function from the equal-styled variety. The
anthers in the six equal-styled and in the apparently twelve
short-styled plants were seated high up in the corolla, as in the
true short-styled cowslip; and the pollen-grains resembled
those of the same form in their large size, but were mingled

with a few shrivelled grains. In function this pollen was identical with that of the short-styled cowslip; for ten long-styled flowers of the common cowslip, legitimately fertilised with pollen from a true equal-styled variety, produced six capsules, containing on an average 34·4 seeds; whilst seven capsules on a short-styled cowslip illegitimately fertilised with pollen from the equal-styled variety, yielded an average of only 14·5 seeds.

As the equal-styled plants differ from one another in their powers of reproduction, and as this is an important subject, I will give a few details with respect to five of them. First, an equal-styled plant, protected from insects (as was done in all the following cases, with one stated exception), spontaneously produced numerous capsules, five of which gave an average of 44·8 seeds, with a maximum in one capsule of 57. But six capsules, the product of fertilisation with pollen from a short-styled cowslip (and this is a legitimate union), gave an average of 28·5 seeds, with a maximum of 49; and this is a much lower average than might have been expected. Secondly, nine capsules from another equal-styled plant, which had not been protected from insects, but probably was self-fertilised, gave an average of 45·2 seeds, with a maximum of 58. Thirdly, another plant which had a very short pistil in 1865, produced spontaneously many capsules, six of which contained an average of 33·9 seeds, with a maximum of 38. In 1866 this same plant had a pistil of wonderful length; for it projected quite above the anthers, and the stigma resembled that of the long-styled form. In this condition it produced spontaneously a vast number of fine capsules, six of which contained almost exactly the same average number as before, viz. 34·3, with a maximum of 38. Four flowers on this plant, legitimately fertilised with pollen from a short-styled cowslip, yielded capsules with an average of 30·2 seeds. Fourthly another short-styled plant spontaneously produced in 1865 an abundance of capsules, ten of which contained an average of 35·6 seeds, with a maximum of 54. In 1866 this same plant had become in all respects long-styled, and ten capsules gave almost exactly the same average as before, viz. 35·1 seeds, with a maximum of 47. Eight flowers on this plant, legitimately fertilised with pollen from a short-styled cowslip, produced six capsules, with the high average of 53 seeds, and the high maximum of 67. Eight flowers were also fertilised with pollen from a long-styled cowslip (this being an illegitimate union), and produced seven capsules, containing an average of 24·4 seeds, with a maximum of 32. The fifth and last plant remained in the same condition during both years: it had a pistil rather longer than that of the true short-styled form, with the stigma smooth, as it ought to be in this form, but abnormal in shape, like a much-elongated inverted cone. It produced spontaneously many capsules, five of which, in 1865, gave an average of only 15·6 seeds; and in 1866 ten capsules still gave an average only a little higher, viz. of 22·1, with a maximum of 30. Sixteen flowers were fertilised with pollen from a long-styled cowslip, and produced 12 capsules, with an average of 24·9 seeds, and a maximum of 42. Eight flowers were fertilised with pollen from a short-styled cowslip, but yielded only two capsules, containing 18 and 23 seeds. Hence this plant, in function and partially in structure, was in an almost exactly intermediate state between the long-styled and short-styled form, but inclining towards the short-styled; and this accounts for the low average of seeds which it produced when spontaneously self-fertilised.

The foregoing five plants thus differ much from one another in the nature of their fertility. In two individuals a great difference in the length of the pistil during two succeeding years made no difference in the number of seeds produced. As all five plants possessed the male organs of the short-styled form in a perfect state, and the female organs of the long-styled form in a more or less complete state, they spontaneously produced a surprising number of capsules, which generally contained a large average of remarkably fine seeds. With ordinary cowslips, *legitimately fertilised*, I once obtained from plants cultivated in the greenhouse the high average, from seven capsules, of 58·7 seeds, with a maximum in one capsule of 87 seeds; but from plants grown out of doors I never obtained a higher average than 41 seeds. Now two of the equal-styled plants, grown out of doors and spontaneously *self-fertilised*, gave averages of 44 and 45 seeds; but this high fertility may perhaps be in part attributed to the stigma receiving pollen from the surrounding anthers at exactly the right period. Two of these plants, fertilised with pollen from a short-styled cowslip (and this in fact is a legitimate union), gave a lower average than when self-fertilised. On the other hand, another plant, when similarly fertilised by a cowslip, yielded the unusually high average of 53 seeds, with a maximum of 67. Lastly, as we have just seen, one of these plants was in

an almost exactly intermediate condition in its female organs
between the long- and short-styled forms, and consequently,
when self-fertilised, yielded a low average of seed. If we add
together all the experiments which I made on the equal-styled
plants, 41 spontaneously self-fertilised capsules (insects having
been excluded) gave an average of 34 seeds, which is exactly the
same number as the parent-plant yielded in Edinburgh. Thirty-
four flowers, fertilised with pollen from the short-styled cowslip
(and this is an analogous union), produced 17 capsules, containing an average of 33·8 seeds. It is a rather singular circumstance, for which I cannot account, that 20 flowers, artificially fertilised on one occasion with pollen from the same plants yielded
only ten capsules, containing the low average of 26·7 seeds.

As bearing on inheritance, it may be added that 72 seedlings were raised from one of the red-flowered, strictly equal-
styled, self-fertilised plants descended from the similarly characterised Edinburgh plant. These 72 plants were therefore grandchildren of the Edinburgh plant, and they all bore,
as in the first generation, red flowers, with the exception of
one plant, which reverted in colour to the common cowslip.
In regard to structure, nine plants were truly long-styled
and had their stamens seated low down in the corolla in the
proper position; the remaining 63 plants were equal-styled,
though the stigma in about a dozen of them stood a little below
the anthers. We thus see that the anomalous combination in the
same flower, of the male and female sexual organs which properly
exist in the two distinct forms, was inherited with much force.
Thirty-six seedlings were also raised from long and short-styled
common cowslips, crossed with pollen from the equal-styled
variety. Of these plants one alone was equal-styled, 20 were
short-styled, but with the pistil in three of them rather too
long, and the remaining 15 were long-styled. In this case we
have an illustration of the difference between simple inheritance
and prepotency of transmission; for the equal-styled variety,
when self-fertilised, transmits its character, as we have just
seen, with much force, but when crossed with the common
cowslip cannot withstand the greater power of transmission
of the latter.

PULMONARIA.

I have little to say on this genus. I obtained seeds of *P. officinalis* from a garden where the long-styled form alone grew,

and raised 11 seedlings, which were all long-styled. These plants were named for me by Dr. Hooker. They differed, as has been shown, from the plants belonging to this species which in Germany were experimented on by Hildebrand;* for he found that the long-styled form was absolutely sterile with its own pollen, whilst my long-styled seedlings and the parent-plants yielded a fair supply of seed when self-fertilised. Plants of the long-styled form of *Pulmonaria angustifolia* were, like Hildebrand's plants, absolutely sterile with their own pollen, so that I could never procure a single seed. On the other hand, the short-styled plants of this species, differently from those of *P. officinalis*, were fertile with their own pollen in a quite remarkable degree for a heterostyled plant. From seeds carefully self-fertilised I raised 18 plants, of which 13 proved short-styled and 5 long-styled.

POLYGONUM FAGOPYRUM.

From flowers on long-styled plants fertilised illegitimately with pollen from the same plant, 49 seedlings were raised, and these consisted of 45 long-styled and 4 short-styled. From flowers on short-styled plants illegitimately fertilised with pollen from the same plant 33 seedlings were raised, and these consisted of 20 short-styled and 13 long-styled. So that the usual rule of illegitimately fertilised long-styled plants tending much more strongly than short-styled plants to reproduce their own form here holds good. The illegitimate plants derived from both forms flowered later than the legitimate, and were to the latter in height as 69 to 100. But as these illegitimate plants were descended from parents fertilised with their own pollen, whilst the legitimate plants were descended from parents crossed with pollen from a distinct individual, it is impossible to know how much of their difference in height and period of flowering, is due to the illegitimate birth of the one set, and how much to the other set being the product of a cross between distinct plants.

Concluding Remarks on the Illegitimate Offspring of Heterostyled Trimorphic and Dimorphic Plants.

It is remarkable how closely and in how many points illegitimate unions between the two or three forms of the

---

* 'Bot. Zeitung,' 1865, p. 13.

same heterostyled species, together with their illegitimate offspring, resemble hybrid unions between distinct species together with their hybrid offspring. In both cases we meet with every degree of sterility, from very slightly lessened fertility to absolute barrenness, when not even a single seed-capsule is produced. In both cases the facility of effecting the first union is much influenced by the conditions to which the plants are exposed.* Both with hybrids and illegitimate plants the innate degree of sterility is highly variable in plants raised from the same mother-plant. In both cases the male organs are more plainly affected than the female; and we often find contabescent anthers enclosing shrivelled and utterly powerless pollen-grains. The more sterile hybrids, as Max Wichura has well shown,† are sometimes much dwarfed in stature, and have so weak a constitution that they are liable to premature death; and we have seen exactly parallel cases with the illegitimate seedlings of Lythrum and Primula. Many hybrids are the most persistent and profuse flowerers, as are some illegitimate plants. When a hybrid is crossed by either pure parent-form, it is notoriously much more fertile than when crossed *inter se* or by another hybrid; so when an illegitimate plant is fertilised by a legitimate plant, it is more fertile than when fertilised *inter se* or by another illegitimate plant. When two species are crossed and they produce numerous seeds, we expect as a general rule that their hybrid offspring will be moderately fertile; but if the parent species produce extremely few seeds, we expect that the hybrids will be very

* This has been remarked by many experimentalists in effecting crosses between distinct species; and in regard to illegitimate unions I have given in the first chapter a striking illustration in the case of *Primula veris*.

† 'Die Bastardbefruchtung im Pflanzenreich,' 1865.

sterile. But there are marked exceptions, as shown
by Gärtner, to these rules. So it is with illegitimate
unions and illegitimate offspring. Thus the mid-
styled form of *Lythrum salicaria*, when illegitimately
fertilised with pollen from the longest stamens of
the short-styled form, produced an unusual number
of seeds; and their illegitimate offspring were not at
all, or hardly at all, sterile. On the other hand, the
illegitimate offspring from the long-styled form, fertilised with pollen from the shortest stamens of the same
form, yielded few seeds, and the illegitimate offspring
thus produced were very sterile; but they were more
sterile than might have been expected relatively to the
difficulty of effecting the union of the parent sexual
elements. No point is more remarkable in regard to
the crossing of species than their unequal reciprocity.
Thus species A will fertilise B with the greatest ease;
but B will not fertilise A after hundreds of trials. We
have exactly the same case with illegitimate unions;
for the mid-styled *Lythrum salicaria* was easily fertilised by pollen from the longest stamens of the short-
styled form, and yielded many seeds; but the latter
form did not yield a single seed when fertilised by the
longest stamens of the mid-styled form.

Another important point is prepotency. Gärtner
has shown that when a species is fertilised with pollen
from another species, if it be afterwards fertilised with
its own pollen, or with that of the same species, this
is so prepotent over the foreign pollen that the effect
of the latter, though placed on the stigma some time
previously, is entirely destroyed. Exactly the same
thing occurs with the two forms of a heterostyled
species. Thus several long-styled flowers of *Primula
veris* were fertilised illegitimately with pollen from
another plant of the same form, and twenty-four hours

R

afterwards legitimately with pollen from a short-styled dark-red polyanthus which is a variety of *P. veris*; and the result was that every one of the thirty seedlings thus raised bore flowers more or less red, showing plainly how prepotent the legitimate pollen from a short-styled plant was over the illegitimate pollen from a long-styled plant.

In all the several foregoing points the parallelism is wonderfully close between the effects of illegitimate and hybrid fertilisation. It is hardly an exaggeration to assert that seedlings from an illegitimately fertilised heterostyled plant are hybrids formed within the limits of one and the same species. This conclusion is important, for we thus learn that the difficulty in sexually uniting two organic forms and the sterility of their offspring, afford no sure criterion of so-called specific distinctness. If any one were to cross two varieties of the same form of Lythrum or Primula for the sake of ascertaining whether they were specifically distinct, and he found that they could be united only with some difficulty, that their offspring were extremely sterile, and that the parents and their offspring resembled in a whole series of relations crossed species and their hybrid offspring, he might maintain that his varieties had been proved to be good and true species; but he would be completely deceived. In the second place, as the forms of the same trimorphic or dimorphic heterostyled species are obviously identical in general structure, with the exception of the reproductive organs, and as they are identical in general constitution (for they live under precisely the same conditions), the sterility of their illegitimate unions and that of their illegitimate offspring, must depend exclusively on the nature of the sexual elements and on their incompatibility for uniting in a particular

manner. And as we have just seen that distinct species
when crossed resemble in a whole series of relations the
forms of the same species when illegitimately united,
we are led to conclude that the sterility of the former
must likewise depend exclusively on the incompatible
nature of their sexual elements, and not on any general
difference in constitution or structure. We are, indeed,
led to this same conclusion by the impossibility of de-
tecting any differences sufficient to account for certain
species crossing with the greatest ease, whilst other
closely allied species cannot be crossed, or can be crossed
only with extreme difficulty. We are led to this conclusion still more forcibly by considering the great
difference which often exists in the facility of crossing
reciprocally the same two species; for it is manifest in
this case that the result must depend on the nature of
the sexual elements, the male element of the one
species acting freely on the female element of the
other, but not so in a reversed direction. And now we
see that this same conclusion is independently and
strongly fortified by the consideration of the illegitimate unions of trimorphic and dimorphic heterostyled
plants. In so complex and obscure a subject as hybridism it is no slight gain to arrive at a definite conclusion, namely, that we must look exclusively to functional differences in the sexual elements, as the cause
of the sterility of species when first crossed and of
their hybrid offspring. It was this consideration which
led me to make the many observations recorded in this
chapter, and which in my opinion make them worthy
of publication.

CHAPTER VI.

CONCLUDING REMARKS ON HETEROSTYLED PLANTS.

The essential character of heterostyled plants—Summary of the differences in fertility between legitimately and illegitimately fertilised plants—Diameter of the pollen-grains, size of anthers and structure of stigma in the different forms—Affinities of the genera which include heterostyled species—Nature of the advantages derived from heterostylism—The means by which plants became heterostyled—Transmission of form—Equal-styled varieties of heterostyled plants—Final remarks.

In the foregoing chapters all the heterostyled plants known to me have been more or less fully described. Several other cases have been indicated, especially by Professor Asa Gray and Kuhn,* in which the individuals of the same species differ in the length of their stamens and pistils; but as I have been often deceived by this character taken alone, it seems to me the more prudent course not to rank any species as heterostyled, unless we have evidence of more important differences between the forms, as in the diameter of the pollen-grains, or in the structure of the stigma. The individuals of many ordinary hermaphrodite plants habitually fertilise one another, owing to their male and female organs being mature at different periods, or to the structure of the parts, or to self-sterility, &c.; and so it is with many hermaphrodite animals, for instance, land-snails or earth-worms; but in all these cases any one individual can fully fertilise or be ferti-

---

* Asa Gray, 'American Journ. of Science,' 1865, p. 101; and elsewhere as already referred to. Kuhn, 'Bot. Zeitung,' 1867, p. 67.

lised by any other individual of the same species. This is not so with heterostyled plants : a long-styled, mid-styled or short-styled plant cannot fully fertilise or be fertilised by any other individual, but only by one belonging to another form. Thus the essential character of plants belonging to the heterostyled class is that the individuals are divided into two or three bodies, like the males and females of diœcious plants or of the higher animals, which exist in approximately equal numbers and are adapted for reciprocal fertilisation. The existence, therefore, of two or three bodies of individuals, differing from one another in the above more important characteristics, offers by itself good evidence that the species is heterostyled. But absolutely conclusive evidence can be derived only from experiments, and by finding that pollen must be applied from the one form to the other in order to ensure complete fertility.

In order to show how much more fertile each form is when legitimately fertilised with pollen from the other form (or in the case of trimorphic species, with the proper pollen from one of the two other forms) than when illegitimately fertilised with its own-form pollen, I will append a Table (33) giving a summary of the results in all the cases hitherto ascertained. The fertility of the unions may be judged by two standards, namely, by the proportion of flowers which, when fertilised in the two methods, yield capsules, and by the average number of seeds per capsule. When there is a dash in the left-hand column opposite to the name of the species, the proportion of the flowers which yielded capsules was not recorded.

The two or three forms of the same heterostyled species do not differ from one another in general habit or foliage, as sometimes, though rarely, happens with

TABLE 33.

Fertility of the Legitimate Unions taken together, compared with that of the Illegitimate Unions together. The fertility of the Legitimate Unions, as judged by both standards, is taken as 100.

Name of Species.	Illegitimate Unions.	
	Proportional Number of Flowers which produced Capsules.	Average Number of Seeds per Capsule.
Primula veris	69	65
P. elatior	27	75
P. vulgaris	60	54
P. Sinensis	84	63
P. Sinensis (second trial)	0	53
P. Sinensis (Hildebrand)	100	42
P. auricula (Scott)	80	15
P. Sikkimensis ,,	95	31
P. cortusoides ,,	74	66
P. involucrata ,,	72	48
P. farinosa ,,	71	44
Average of the nine species of Primula	88·4	69
Hottonia palustris (H. Müller)	—	61
Linum grandiflorum (the difference probably is much greater)	—	69
L. perenne	—	20
L. perenne (Hildebrand)	0	0
Pulmonaria officinalis (German stock, Hildebrand)	0	0
Pulmonaria angustifolia	35	32
Mitchella repens	20	47
Borreria, Brazilian sp.	—	0
Polygonum fagopyrum	—	46
Lythrum salicaria	33	46
Oxalis Valdiviana (Hildebrand)	2	34
O. Regnelli ,,	0	0
O. speciosa	15	49

the two sexes of dicecious plants. Nor does the calyx differ, but the corolla sometimes differs slightly in shape, owing to the different position of the anthers. In Borreria the hairs within the tube of the corolla are differently situated in the long-styled and short-styled forms. In Pulmonaria there is a slight difference in the size of

the corolla, and in Pontederia in its colour. In the reproductive organs the differences are much greater and
more important. In the one form the stamens may be
all of the same length, and in the other graduated in
length, or alternately longer and shorter. The filaments may differ in colour and thickness, and are
sometimes nearly thrice as long in the one form as in the
other. They adhere also for very different proportional
lengths to the corolla.   The anthers sometimes differ
much in size in the two forms.   Owing to the rotation
of the filaments, the anthers, when mature, dehisce towards the circumference of the flower in one form of
Faramea, and towards the centre in the other form. The
pollen-grains sometimes differ conspicuously in colour,
and often to an extraordinary degree in diameter.
They differ also somewhat in shape, and apparently in
their contents, as they are unequally opaque.   In the
short-styled form of Faramea the pollen-grains are
covered with sharp points, so as to cohere readily together or to an insect; whilst the smaller grains of the
long-styled form are quite smooth.

With respect to the pistil, the style may be almost
thrice as long in the one form as in the other.   In
Oxalis it sometimes differs in hairiness in the three
forms.   In Linum the pistils either diverge and pass
out between the filaments, or stand nearly upright and
parallel to them.   The stigmas in the two forms often
differ much in size and shape, and more especially in
the length and thickness of their papillæ; so that
the surface may be rough or quite smooth.   Owing to
the rotation of the styles, the papillose surface of
the stigma is turned outwards in one form of *Linum
perenne,* and inwards in the other form.   In flowers of
the same age of *Primula veris* the ovules are larger in
the long-styled than in the short-styled form.   The

seeds produced by the two or three forms often differ in number, and sometimes in size and weight; thus, five seeds from the long-styled form of *Lythrum salicaria* equal in weight six from the mid-styled and seven from the short-styled form. Lastly, short-styled plants of *Pulmonaria officinalis* bear a larger number of flowers, and these set a larger proportional number of fruit, which however yield a lower average number of seed, than the long-styled plants. With heterostyled plants we thus see in how many and in what important characters the forms of the same undoubted species often differ from one another—characters which with ordinary plants would be amply sufficient to distinguish species of the same genus.

As the pollen-grains of ordinary species belonging to the same genus generally resemble one another closely in all respects, it is worth while to show, in the following table (34), the difference in diameter between the grains from the two or three forms of the same heterostyled species in the forty-three cases in which this was ascertained. But it should be observed that some of the following measurements are only approximately accurate, as only a few grains were measured. In several cases, also, the grains had been dried and were then soaked in water. Whenever they were of an elongated shape their longer diameters were measured. The grains from the short-styled plants are invariably larger than those from the long-styled, whenever there is any difference between them. The diameter of the former is represented in the table by the number 100.

We here see that, with seven or eight exceptions out of the forty-three cases, the pollen-grains from one form are larger than those from the other form of the same species. The extreme difference is as 100 to 55;

TABLE 34.

Relative Diameter of the Pollen-grains from the forms of the same Heterostyled Species; those from the short-styled form being represented by 100.

Dimorphic Species.

	From the Long-styled form		From the Long-styled form.
Primula veris	67	Cordia (sp. ?)	100
„ vulgaris	71	Gilia pulchella	100
„ Sinensis(Hildebrand)	57	„ micrantha	81
„ auricula	71	Sethia acuminata	83
Hottonia palustris (H. Müller)	61	Erythroxylum (sp. ?)	93
„ „ (self)	64	Cratoxylon formosum	86
Linum grandiflorum	100	Mitchella repens, pollen-grains of the long-styled a little smaller.	
„ perenne(diameter variable). }	100 (?)		
„ flavum	100	Borreria (sp. ?)	92
Pulmonaria officinalis	78	Faramea (sp. ?)	67
„ angustifolia	91	Suteria (sp. ?) (Fritz Müller)	75
Polygonum fagopyrum	82	Houstonia cœrulea	72
Leucosmia Burnettiana	99	Oldenlandia (sp. ?)	78
Ægiphila elata	62	Hedyotis (sp. ?)	88
Menyanthes trifoliata	84	Coccocypselum (sp.?) (F. Müller) }	100
Limnanthemum Indicum	100		
Villarsia (sp. ?)	75	Lipostoma (sp. ?)	80
Forsythia suspensa	94	Cinchona micrantha	91

Trimorphic Species.

Ratio expressing the extreme differences in diameter of the pollen-grains from the two sets of anthers in the three forms.		Ratio between the diameters of the pollen-grains of the two sets of anthers in the same form.	
Lythrum salicaria	60	Oxalis rosea, long-styled form (Hildebrand). }	83
Nesæa verticillata	65		
Oxalis Valdiviana (Hildebrand)	71	„ compressa, short-styled form }	83
„ Regnelli	78		
„ speciosa	69	Pontederia (sp. ?) short-styled form. }	87
„ sensitiva	84		
Pontederia (sp. ?)	55	„ other sp., mid-styled form. }	86

and we should bear in mind that in the case of spheres differing to this degree in diameter, their contents differ in the ratio of six to one. With all the species in which the grains differ in diameter, there is no exception to the rule that those from the anthers of the short-styled form, the tubes of which have to penetrate the longer pistil of the long-styled form, are larger than the grains from the other form. This curious relation led Delpino* (as it formerly did me) to believe that the larger size of the grains in the short-styled flowers is connected with the greater supply of matter needed for the development of their longer tubes. But the case of Linum, in which the grains of the two forms are of equal size, whilst the pistil of the one is about twice as long as that of the other, made me from the first feel very doubtful with respect to this view. My doubts have since been strengthened by the cases of Limnanthemum and Coccocypselum, in which the grains are of equal size in the two forms; whilst in the former genus the pistil is nearly thrice and in the latter twice as long as in the other form. In those species in which the grains are of unequal size in the two forms, there is no close relationship between the degree of their inequality and that of their pistils. Thus in *Pulmonaria officinalis* and in Erythroxylum the pistil in the long-styled form is about twice the length of that in the other form, whilst in the former species the pollen-grains are as 100 to 78, and in the latter as 100 to 93 in diameter. In the two forms of Suteria the pistil differs but little in length, whilst the pollen - grains are as 100 to 75 in diameter. These cases seem to prove that the difference in size between the grains in the two forms is not determined by the length of the pistil, down which the tubes have to grow. That with plants in general there is no close relationship between

* 'Sull' Opera, la Distribuzione dei Sessi nelle Piante,' &c., 1867, p. 17.

the size of the pollen-grains and the length of the pistil is manifest: for instance, I found that the distended grains of *Datura arborea* were ·00243 of an inch in diameter, and the pistil no less than 9·25 inches in length; now the pistil in the small flowers of *Polygonum fagopyrum* is very short, yet the larger pollen-grains from the short-styled plants had exactly the same diameter as those from the Datura, with its enormously elongated pistil.

Notwithstanding these several considerations, it is difficult quite to give up the belief that the pollen-grains from the longer stamens of heterostyled plants have become larger in order to allow of the development of longer tubes; and the foregoing opposing facts may possibly be reconciled in the following manner. The tubes are at first developed from matter contained within the grains, for they are sometimes exserted to a considerable length, before the grains have touched the stigma; but botanists believe that they afterwards draw nourishment from the conducting tissue of the pistil. It is hardly possible to doubt that this must occur in such cases as that of the Datura, in which the tubes have to grow down the whole length of the pistil, and therefore to a length equalling 3,806 times the diameter of the grains (namely, ·00243 of an inch) from which they are protruded. I may here remark that I have seen the pollen-grains of a willow, immersed in a very weak solution of honey, protrude their tubes, in the course of twelve hours, to a length thirteen times as great as the diameter of the grains. Now if we suppose that the tubes in some heterostyled species are developed wholly or almost wholly from matter contained within the grains, while in other species from matter yielded by the pistil, we can see that in the former case it would be necessary that the grains of the two forms should differ in size relatively to the length of the pistil which the tubes have to penetrate, but that in the latter case it would not be necessary that the grains should thus differ. Whether this explanation can be considered satisfactory must remain at present doubtful.

There is another remarkable difference between the forms of several heterostyled species, namely in the anthers of the short-styled flowers, which contain the larger pollen-grains, being longer than those of the long-styled flowers. This is the case with *Hottonia palustris* in the ratio of 100 to 83. With *Limnanthemum Indicum* the ratio is as 100 to 70. With the allied Menyanthes the anthers of the short-styled form are a little and with Villarsia conspicuously larger than those of the long-styled. With *Pulmonaria angustifolia* they vary much in size, but from an average of seven measurements of each kind the ratio is as 100 to 91. In six genera of the Rubiaceæ there is a similar difference, either slightly or well marked. Lastly, in the trimorphic Pontederia the ratio is 100 to 88; the anthers from the longest stamens in the short-styled form being compared with those from the shortest stamens in the long-styled form. On the other hand, there is a similar and well-marked difference in the length of the stamens in the two forms of *Forsythia suspensa* and of *Linum flavum;* but in these two cases the anthers of the short-styled flowers are shorter than those of the long-styled. The relative size of the anthers was not particularly attended to in the two forms of the other heterostyled plants, but I believe that they are generally equal, as is certainly the case with those of the common primrose and cowslip.

The pistil differs in length in the two forms of every

heterostyled plant, and although a similar difference
is very general with the stamens, yet in the two
forms of *Linum grandiflorum* and of Cordia they are
equal. There can hardly be a doubt that the relative length of these organs is an adaptation for the
safe transportal by insects of the pollen from the one
form to the other. The exceptional cases in which
these organs do not stand exactly on a level in the two
forms may probably be explained by the manner in
which the flowers are visited. With most of the
species, if there is any difference in the size of the
stigma in the two forms, that of the long-styled, whatever its shape may be, is larger than that of the short-
styled. But here again there are some exceptions to
the rule, for in the short-styled form of *Leucosmia
Burnettiana* the stigmas are longer and much narrower
than those of the long-styled; the ratio between the
lengths of the stigmas in the two forms being 100 to 60.
In the three Rubiaceous genera, Faramea, Houstonia
and Oldenlandia, the stigmas of the short-styled form
are likewise somewhat longer and narrower; and in
the three forms of *Oxalis sensitiva* the difference is
strongly marked, for if the length of the two stigmas
of the long-styled pistil be taken as 100, it will be
represented in the mid- and short-styled forms by
the numbers 141 and 164. As in all these cases the
stigmas of the short-styled pistil are seated low down
within a more or less tubular corolla, it is probable
that they are better fitted by being long and narrow
for brushing the pollen off the inserted proboscis of
an insect.

With many heterostyled plants the stigma differs
in roughness in the two forms, and when this is the
case there is no known exception to the rule that the
papillæ on the stigma of the long-styled form are longer

and often thicker than those on that of the short-styled. For instance, the papillæ on the long-styled stigma of *Hottonia palustris* are more than twice the length of those in the other form. This holds good even in the case of *Houstonia cœrulea*, in which the stigmas are much shorter and stouter in the long-styled than in the short-styled form, for the papillæ on the former compared with those on the latter are as 100 to 58 in length. The length of the pistil in the long-styled form of *Linum grandiflorum* varies much, and the stigmatic papillæ vary in a corresponding manner. From this fact I inferred at first that in all cases the difference in length between the stigmatic papillæ in the two forms was one merely of correlated growth; but this can hardly be the true or general explanation, as the shorter stigmas of the long-styled form of Houstonia have the longer papillæ. It is a more probable view that the papillæ, which render the stigma of the long-styled form of various species rough, serve to entangle effectually the large-sized pollen-grains brought by insects from the short-styled form, thus ensuring its legitimate fertilisation. This view is supported by the fact that the pollen-grains from the two forms of eight species in Table 34 hardly differ in diameter, and the papillæ on their stigmas do not differ in length.

The species which are at present positively or almost positively known to be heterostyled belong, as shown in the following table, to 38 genera, widely distributed throughout the world. These genera are included in fourteen Families, most of which are very distinct from one another, for they belong to nine of the several great Series, into which phanerogamic plants have been divided by Bentham and Hooker.

TABLE 35.

List of Genera including Heterostyled Species.

DICOTYLEDONS.		DICOTYLEDONS.	
Cratoxylon.	Hypericineæ.	Mitchella.	Rubiaceæ.
Erythroxylum.	Erythroxyleæ.	Diodia.	,,
Sethia.	,,	Borreria.	,,
Linum.	Geraniaceæ.	Spermacoce.	,,
Oxalis.	,,	Primula.	Primulaceæ.
Lythrum.	Lythraceæ.	Hottonia.	,,
Nesæa.	,,	Androsace.	,,
Cinchona.	Rubiaceæ.	Forsythia.	Oleaceæ.
Bouvardia.	,,	Menyanthes.	Gentianaceæ.
Manettia.	,,	Limnanthemum.	,,
Hedyotis.	,,	Villarsia.	,,
Oldenlandia.	,,	Gilia.	Polemoniaceæ.
Houstonia.	,,	Cordia.	Cordieæ.
Coccocypselum.	,,	Pulmonaria.	Boragineæ.
Lipostoma.	,,	Ægiphila.	Verbenaceæ.
Knoxia.	,,	Polygonum.	Polygoneæ.
Faramea.	,,	Thymelea.	Thymeleæ.
Psychotria.	,,		
Rudgea.	,,	MONOCOTYLEDONS.	
Suteria.	,,	Pontederia.	Pontederiaceæ.

In some of these families the heterostyled condition must have been acquired at a very remote period. Thus the three closely allied genera, Menyanthes, Limnanthemum, and Villarsia, inhabit respectively Europe, India, and South America. Heterostyled species of Hedyotis are found in the temperate regions of North and the tropical regions of South America. Trimorphic species of Oxalis live on both sides of the Cordillera in South America and at the Cape of Good Hope. In these and some other cases it is not probable that each species acquired its heterostyled structure independently of its close allies. If they did not do so, the three closely connected genera of the Menyantheæ and the several trimorphic species of Oxalis must have inherited their structure from a common progenitor. But an immense lapse of time will have been necessary in all such cases for the modified descendants of a common progenitor to have

spread from a single centre to such widely remote and
separated areas. The family of the Rubiaceæ contains
not far short of as many heterostyled genera as all
the other thirteen families together; and hereafter
no doubt other Rubiaceous genera will be found to
be heterostyled, although a large majority are homostyled. Several closely allied genera in this family
probably owe their heterostyled structure to descent
in common; but as the genera thus characterised are
distributed in no less than eight of the tribes into
which this family has been divided by Bentham and
Hooker, it is almost certain that several of them
must have become heterostyled independently of
one another. What there is in the constitution or
structure of the members of this family which favours
their becoming heterostyled, I cannot conjecture.
Some families of considerable size, such as the Boragineæ and Verbenaceæ, include, as far as is at
present known, only a single heterostyled genus.
Polygonum also is the sole heterostyled genus in its
family; and though it is a very large genus, no other
species except P. fagopyrum is thus characterised. We
may suspect that it has become heterostyled within
a comparatively recent period, as it seems to be less
strongly so in function than the species in any other
genus, for both forms are capable of yielding a considerable number of spontaneously self-fertilised seeds.
Polygonum in possessing only a single heterostyled
species is an extreme case; but every other genus of
considerable size which includes some such species
likewise contains homostyled species. Lythrum includes trimorphic, dimorphic, and homostyled species.

Trees, bushes, and herbaceous plants, both large
and small, bearing single flowers or flowers in dense
spikes or heads, have been rendered heterostyled.

So have plants which inhabit alpine and lowland sites, dry land, marshes and water.*

When I first began to experimentise on hetero-styled plants it was under the impression that they were tending to become diœcious; but I was soon forced to relinquish this notion, as the long-styled plants of Primula which, from possessing a longer pistil, larger stigma, shorter stamens with smaller pollen-grains, seemed to be the more feminine of the two forms, yielded fewer seeds than the short-styled plants which appeared to be in the above respects the more mascu-line of the two. Moreover, trimorphic plants evidently come under the same category with dimorphic, and the former cannot be looked at as tending to become diœcious. With *Lythrum salicaria,* however, we have the curious and unique case of the mid-styled form being more feminine or less masculine in nature than the other two forms. This is shown by the large

* Out of the 38 genera known to include heterostyled species, about eight, or 21 per cent., are more or less aquatic in their habits. I was at first struck with this fact, for I was not then aware how large a proportion of or-dinary plants inhabit such sta-tions. Heterostyled plants may be said in one sense to have their sexes separated, as the forms must mutually fertilise one another. Therefore it seemed worth while to ascertain what proportion of the genera in the Linnean classes, Monœcia, Diœcia and Poly-gamia, contained species which live "in water, marshes, bogs or watery places." In Sir W. J. Hooker's 'British Flora' (4th edit. 1838) these three Linnean classes include 40 genera, 17 of which (i.e. 43 per cent.) contain species inhabiting the just-speci-fied stations. So that 43 per cent. of those British plants which have their sexes separated are more or less aquatic in their habits, whereas only 21 per cent. of heterostyled plants have such habits. I may add that the her-maphrodite classes, from Monan-dria to Gynandria inclusive, con-tain 447 genera, of which 113 are aquatic in the above sense, or only 25 per cent. It thus appears, as far as can be judged from such imperfect data, that there is some connection between the separation of the sexes in plants and the watery nature of the sites which they inhabit; but that this does not hold good with heterostyled species.

number of seeds which it yields in whatever manner
it may be fertilised, and by its pollen (the grains of
which are of smaller size than those from the corresponding stamens in the other two forms) when
applied to the stigma of any form producing fewer
seeds than the normal number. If we suppose the
process of deterioration of the male organs in the mid-
styled form to continue, the final result would be the
production of a female plant; and *Lythrum salicaria*
would then consist of two heterostyled hermaphrodites
and a female. No such case is known to exist, but it
is a possible one, as hermaphrodite and female forms
of the same species are by no means rare. Although
there is no reason to believe that heterostyled plants
are regularly becoming diœcious, yet they offer singular facilities, as will hereafter be shown, for such
conversion; and this appears occasionally to have been
effected.

We may feel sure that plants have been rendered
heterostyled to ensure cross-fertilisation, for we now
know that a cross between the distinct individuals of
the same species is highly important for the vigour and
fertility of the offspring. The same end is gained by
dichogamy or the maturation of the reproductive elements of the same flower at different periods,—by
diœciousness—self-sterility—the prepotency of pollen
from another individual over a plant's own pollen,—and
lastly, by the structure of the flower in relation to the
visits of insects. The wonderful diversity of the means
for gaining the same end in this case, and in many
others, depends on the nature of all the previous
changes through which the species has passed, and on
the more or less complete inheritance of the successive
adaptations of each part to the surrounding conditions.

Plants which are already well adapted by the structure
of their flowers for cross-fertilisation by the aid of
insects often possess an irregular corolla, which has
been modelled in relation to their visits ; and it would
have been of little or no use to such plants to have
become heterostyled. We can thus understand why
it is that not a single species is heterostyled in such
great families as the Leguminosæ, Labiatæ, Scrophu-
lariaceæ, Orchideæ, &c., all of which have irregular
flowers. Every known heterostyled plant, however,
depends on insects for its fertilisation, and not on the
wind ; so that it is a rather surprising fact that only
one genus, Pontederia, has a plainly irregular corolla.

Why some species are adapted for cross-fertilisation,
whilst others within the same genus are not so, or
if they once were, have since lost such adaptation
and in consequence are now usually self-fertilised, I
have endeavoured elsewhere to explain to a certain
limited extent.* If it be further asked why some
species have been adapted for this end by being made
heterostyled, rather than by any of the above specified
means, the answer probably lies in the manner in
which heterostylism originated,—a subject immediately to be discussed. Heterostyled species, however,
have an advantage over dichogamous species, as all
the flowers on the same heterostyled plant belong to
the same form, so that when fertilised legitimately by
insects two distinct individuals are sure to intercross.
On the other hand, with dichogamous plants, early or
late flowers on the same individual may intercross ;
and a cross of this kind does hardly any or no good.
Whenever it is profitable to a species to produce a

---

* 'The Effects of Cross and Self-fertilisation,' 1876, p. 441.

large number of seeds and this obviously is a very
common case, heterostyled will have an advantage
over diœcious plants, as all the individuals of the
former, whilst only half of the latter, that is the
females, yield seeds.   On the other hand, heterostyled plants seem to have no advantage, as far as
cross-fertilisation is concerned, over those which are
sterile with their own pollen.   They lie indeed under
a slight disadvantage, for if two self-sterile plants
grow near together and far removed from all other
plants of the same species, they will mutually and
perfectly fertilise one another, whilst this will not be
the case with heterostyled dimorphic plants, unless
they chance to belong to opposite forms.

It may be added that species which are trimorphic
have one slight advantage over the dimorphic; for if
only two individuals of a dimorphic species happen
to grow near together in an isolated spot, the chances
are even that both will belong to the same form, and
in this case they will not produce the full number of
vigorous and fertile seedlings; all these, moreover,
will tend strongly to belong to the same form as their
parents.   On the other hand, if two plants of the same
trimorphic species happen to grow in an isolated spot,
the chances are two to one in favour of their not belonging to the same form; and in this case they will
legitimately fertilise one another, and yield the full
complement of vigorous offspring.

The Means by which Plants may have been rendered Heterostyled.

This is a very obscure subject, on which I can throw
little light, but which is worthy of discussion.   It has

been shown that heterostyled plants occur in fourteen
natural families, dispersed throughout the whole vegetable kingdom, and that even within the family of the
Rubiaceæ they are dispersed in eight of the tribes. We
may therefore conclude that this structure has been
acquired by various plants independently of inheritance
from a common progenitor, and that it can be acquired
without any great difficulty—that is, without any very
unusual combination of circumstances.

It is probable that the first step towards a species
becoming heterostyled is great variability in the length
of the pistil and stamens, or of the pistil alone. Such
variations are not very rare : with *Amsinckia spectabilis*
and *Nolana prostrata* these organs differ so much in
length in different individuals that, until experiment-
ing on them, I thought both species heterostyled.
The stigma of *Gesneria pendulina* sometimes protrudes
far beyond, and is sometimes seated beneath the
anthers ; so it is with *Oxalis acetosella* and various
other plants. I have also noticed an extraordinary
amount of difference in the length of the pistil in cultivated varieties of *Primula veris* and *vulgaris*.

As most plants are at least occasionally cross-fertilised by the aid of insects, we may assume that this
was the case with our supposed varying plant; but
that it would have been beneficial to it to have been
more regularly cross-fertilised. We should bear in
mind how important an advantage it has been
proved to be to many plants, though in different
degrees and ways, to be cross-fertilised. It might
well happen that our supposed species did not vary
in function in the right manner, so as to become
either dichogamous or completely self-sterile, or in
structure so as to ensure cross-fertilisation. If it had

thus varied, it would never have been rendered heterostyled, as this state would then have been superfluous.
But the parent-species of our several existing heterostyled plants may have been, and probably were (judging from their present constitution) in some degree
self-sterile ; and this would have made regular cross-
fertilisation still more desirable.

Now let us take a highly varying species with most
or all of the anthers exserted in some individuals, and
in others seated low down in the corolla ; with the
stigma also varying in position in like manner. Insects
which visited such flowers would have different parts
of their bodies dusted with pollen, and it would be a
mere chance whether this were left on the stigma of
the next flower which was visited. If all the anthers
could have been placed on the same level in all the
plants, then abundant pollen would have adhered to
the same part of the body of the insects which frequented the flowers, and would afterwards have been
deposited without loss on the stigma, if it likewise
stood on the same unvarying level in all the flowers.
But as the stamens and pistils are supposed to have
already varied much in length and to be still varying,
it might well happen that they could be reduced much
more easily through natural selection into two sets of
different lengths in different individuals, than all to
the same length and level in all the individuals. We
know from innumerable instances, in which the two
sexes and the young of the same species differ, that
there is no difficulty in two or more sets of individuals
being formed which inherit different characters. In
our particular case the law of compensation or balancement (which is admitted by many botanists) would
tend to cause the pistil to be reduced in those indi-

viduals in which the stamens were greatly developed, and to be increased in length in those which had their stamens but little developed.

Now if in our varying species the longer stamens were to be nearly equalised in length in a considerable body of individuals, with the pistil more or less reduced; and in another body, the shorter stamens to be similarly equalised, with the pistil more or less increased in length, cross-fertilisation would be secured with little loss of pollen; and this change would be so highly beneficial to the species, that there is no difficulty in believing that it could be effected through natural selection. Our plant would then make a close approach in structure to a heterostyled dimorphic species; or to a trimorphic species, if the stamens were reduced to two lengths in the same flower in correspondence with that of the pistils in the other two forms. But we have not as yet even touched on the chief difficulty in understanding how heterostyled species could have originated. A completely self-sterile plant or a dichogamous one can fertilise and be fertilised by any other individual of the same species; whereas the essential character of a heterostyled plant is that an individual of one form cannot fully fertilise or be fertilised by an individual of the same form, but only by one belonging to another form.

H. Müller has suggested * that ordinary or homostyled plants may have been rendered heterostyled merely through the effects of habit. Whenever pollen from one set of anthers is habitually applied to a pistil of particular length in a varying species, he believes that at last the possibility of fertilisation in any other

---

* 'Die Befruchtung der Blumen,' p. 352.

manner will be nearly or completely lost. He was led to this view by observing that Diptera frequently carried pollen from the long-styled flowers of Hottonia to the stigma of the same form, and that this illegitimate union was not nearly so sterile as the corresponding union in other heterostyled species. But this conclusion is directly opposed by some other cases, for instance by that of *Linum grandiflorum*; for here the long-styled form is utterly barren with its own-form pollen, although from the position of the anthers this pollen is invariably applied to the stigma. It is obvious that with heterostyled dimorphic plants the two female and the two male organs differ in power; for if the same kind of pollen be placed on the stigmas of the two forms, and again if the two kinds of pollen be placed on the stigmas of the same form, the results are in each case widely different. Nor can we see how this differentiation of the two female and two male organs could have been effected merely through each kind of pollen being habitually placed on one of the two stigmas.

Another view seems at first sight probable, namely, that an incapacity to be fertilised in certain ways has been specially acquired by heterostyled plants. We may suppose that our varying species was somewhat sterile (as is often the case) with pollen from its own stamens, whether these were long or short; and that such sterility was transferred to all the individuals with pistils and stamens of the same length, so that these became incapable of intercrossing freely; but that such sterility was eliminated in the case of the individuals which differed in the length of their pistils and stamens. It is, however, incredible that so peculiar a form of mutual infertility should have been specially

acquired unless it were highly beneficial to the species;
and although it may be beneficial to an individual
plant to be sterile with its own pollen, cross-fertilisation being thus ensured, how can it be any advantage to a plant to be sterile with half its brethren,
that is, with all the individuals belonging to the
same form? Moreover, if the sterility of the unions
between plants of the same form had been a special
acquirement, we might have expected that the long-
styled form fertilised by the long-styled would have
been sterile in the same degree as the short-styled
fertilised by the short-styled; but this is hardly ever
the case. On the contrary, there is sometimes the
widest difference in this respect, as between the two
illegitimate unions of *Pulmonaria angustifolia* and of
Hottonia palustris.

It is a more probable view that the male and female
organs in two sets of individuals have been by some
means specially adapted for reciprocal action; and
that the sterility between the individuals of the same
set or form is an incidental and purposeless result.
The meaning of the term "incidental" may be illustrated by the greater or less difficulty in grafting or
budding together two plants belonging to distinct
species; for as this capacity is quite immaterial to the
welfare of either, it cannot have been specially acquired, and must be the incidental result of differences in their vegetative systems. But how the
sexual elements of heterostyled plants came to differ
from what they were whilst the species was homostyled, and how they became co-adapted in two sets of
individuals, are very obscure points. We know that
in the two forms of our existing heterostyled plants
the pistil always differs, and the stamens generally
differ in length; so does the stigma in structure,

the anthers in size, and the pollen-grains in diameter.
It appears, therefore, at first sight probable that
organs which differ in such important respects could
act on one another only in some manner for which
they had been specially adapted. The probability of
this view is supported by the curious rule that the
greater the difference in length between the pistils
and stamens of the trimorphic species of Lythrum and
Oxalis, the products of which are united for reproduction, by so much the greater is the infertility of the
union. The same rule applies to the two illegitimate
unions of some dimorphic species, namely, *Primula
vulgaris* and *Pulmonaria angustifolia;* but it entirely
fails in other cases, as with *Hottonia palustris* and
Linum grandiflorum. We shall, however, best perceive
the difficulty of understanding the nature and origin
of the co-adaptation between the reproductive organs
of the two forms of heterostyled plants, by considering the case of *Linum grandiflorum*: the two forms of
this plant differ exclusively, as far as we can see, in
the length of their pistils · ·n the long-styled form,
the stamens equal the pistil in length, but their
pollen has no more effect on it than so much inorganic dust; whilst this pollen fully fertilises the
short pistil of the other form. Now, it is scarcely
credible that a mere difference in the length of the
pistil can make a wide difference in its capacity for
being fertilised. We can believe this the less because
with some plants, for instance, *Amsinckia spectabilis*,
the pistil varies greatly in length without affecting
the fertility of the individuals which are intercrossed.
So again I observed that the same plants of *Primula
veris* and *vulgaris* differed to an extraordinary degree
in the length of their pistils during successive seasons;
nevertheless they yielded during these seasons exactly

the same average number of seeds when left to fertilise themselves spontaneously under a net.

We must therefore look to the appearance of inner or hidden constitutional differences between the individuals of a varying species, of such a nature that the male element of one set is enabled to act efficiently only on the female element of another set. We need not doubt about the possibility of variations in the constitution of the reproductive system of a plant, for we know that some species vary so as to be completely self-sterile or completely self-fertile, either in an apparently spontaneous manner or from slightly changed conditions of life. Gärtner also has shown* that the individual plants of the same species vary in their sexual powers in such a manner that one will unite with a distinct species much more readily than another. But what the nature of the inner constitutional differences may be between the sets or forms of the same varying species, or between distinct species, is quite unknown. It seems therefore probable that the species which have become heterostyled at first varied so that two or three sets of individuals were formed differing in the length of their pistils and stamens and in other co-adapted characters, and that almost simultaneously their reproductive powers became modified in such a manner that the sexual elements in one set were adapted to act on the sexual elements of another set; and consequently that these elements in the same set or form incidentally became ill-adapted for mutual interaction, as in the case of distinct species. I have elsewhere shown † that the sterility of species when

---

* Gärtner, 'Bastarderzeugung im Pflanzenreich,' 1849, p. 165.

† 'Origin of Species,' 6th edit. p. 247; 'Variation of Animals and Plants under Domestication,' 2nd edit. vol. ii. p. 169; 'The Effects of Cross and Self-fertilisation,' p.463. It may be well here to remark

first crossed and of their hybrid offspring must also be looked at as merely an incidental result, following from the special co-adaptation of the sexual elements of the same species. We can thus understand the striking parallelism, which has been shown to exist between the effects of illegitimately uniting hetero-styled plants and of crossing distinct species. The great difference in the degree of sterility between the various heterostyled species when illegitimately fer-tilised, and between the two forms of the same species when similarly fertilised, harmonises well with the view that the result is an incidental one which follows from changes gradually effected in their reproductive systems, in order that the sexual elements of the dis-tinct forms should act perfectly on one another.

Transmission of the Two Forms by Heterostyled Plants.—The transmission of the two forms by heterostyled plants, with respect to which many facts were given in the last chapter, may perhaps be found hereafter to throw some light on their manner of development. Hildebrand observed that seedlings from the long-styled form of *Primula Sinensis* when fertilised with pollen from the same form were mostly long-styled, and many analogous cases have since been observed by me. All the known cases are given in the two following tables.

---

that, judging from the remark-able power with which abruptly changed conditions of life act on the reproductive system of most organisms, it is probable that the close adaptation of the male to the female elements in the two forms of the same heterostyled species, or in all the individuals of the same ordinary species, could be acquired only under long-continued nearly uniform conditions of life.

TABLE 36.

Nature of the Offspring from Illegitimately fertilised Dimorphic Plants.

		Number of Long-styled Offspring.	Number of Short-styled Offspring.
Primula veris . .	Long-styled form, fertilised by own-form pollen during five successive generations, produced.	156	6
„ „ . .	Short-styled form, fertilised by own-form pollen, produced .	5	9
Primula vulgaris .	Long-styled form, fertilised by own-form pollen during two successive generations, produced.	69	0
Primula auricula .	Short-styled form, fertilised by own-form pollen, is said to produce during successive generations offspring in about the following proportions .	25	75
Primula Sinensis . .	Long-styled form, fertilised by own-form pollen during two successive generations, produced.	52	0
„ „ . .	Long-styled form, fertilised by own-form pollen (Hildebrand), produced. . . .	14	3
„ „ . .	Short-styled form, fertilised by own-form pollen, produced .	1	24
Pulmonaria officinalis	Long-styled form, fertilised by own-form pollen, produced .	11	0
Polygonum fagopyrum	Long-styled form, fertilised by own-form pollen, produced .	45	4
„ „	Short-styled form, fertilised by own-form pollen, produced .	13	20

TABLE 37.

Nature of the Offspring from Illegitimately fertilised Trimorphic Plants.

—		Number of Long-styled Offspring.	Number of Mid-styled Offspring.	Number of Short-styled Offspring.
Lythrum salicaria.	Long-styled form, fertilised by own-form pollen, produced	56	0	0
,, ,, .	Short-styled form, fertilised by own-form pollen, produced	1	0	8
,, ,, .	Short-styled form, fertilised by pollen from mid-length stamens of long-styled form, produced . . .	4	0	8
,, ,, .	Mid-styled form, fertilised by own-form pollen, produced	1	3	0
,, ,, .	Mid-styled form, fertilised by pollen from shortest stamens of long-styled form, produced	17	8	0
,, ,, .	Mid-styled form, fertilised by pollen from longest stamens of short-styled form, produced	14	8	18
Oxalis rosea . .	Long-styled form, fertilised during several generations by own-form pollen, produced offspring in the ratio of	100	0	0
,, hedysaroides	Mid-styled form, fertilised by own-form pollen, produced	0	17	0

We see in these two tables that the offspring from a form illegitimately fertilised with pollen from another plant of the same form belong, with a few exceptions, to the same form as their parents. For instance, out of 162 seedlings from long-styled plants of *Primula veris* fertilised during five generations in this manner, 156 were long-styled and only 6 short-styled. Of 69 seedlings from *P. vulgaris* similarly raised all were long-styled. So it was with 56 seedlings from the long-styled form of the trimorphic *Lythrum salicaria*, and with numerous seedlings from the long-styled form of *Oxalis rosea*. The offspring from the short-styled forms of dimorphic plants, and from both the mid-styled and short-styled forms of trimorphic plants, fertilised with their own-form pollen, likewise tend to belong to the same form as their parents, but not in so marked a manner as in the case of the long-styled form. There are three cases in Table 37, in which a form of Lythrum was fertilised illegitimately with pollen from another form; and in two of these cases all the offspring belonged to the same two forms as their parents, whilst in the third case they belonged to all three forms.

The cases hitherto given relate to illegitimate unions, but Hildebrand, Fritz Müller, and myself found that a very large proportion, or all of the offspring, from a legitimate union between any two forms of the trimorphic species of Oxalis belonged to the same two forms. A similar rule therefore holds good with unions which are fully fertile, as with those of an illegitimate nature which are more or less sterile. When some of the seedlings from a heterostyled plant belong to a different form from that of its parents, Hildebrand accounts for the fact by reversion. For instance, the long-styled parent-plant of *Primula veris*, from which the 162 illegitimate seedlings in Table 36 were derived in the course of five generations, was itself no doubt derived from the union of a long-styled and a short-styled parent; and the 6 short-styled seedlings may be attributed to reversion to their short-styled progenitor. But it is a surprising fact in this case, and in other similar ones, that the number of the offspring which thus reverted was not larger. The fact is rendered still more strange in the particular instance of *P. veris,* for there was no reversion until four or five generations of long-styled plants had been raised. It may be seen in both tables that the long-styled form transmits its form much more faithfully than does the short-styled, when both are fertilised with their own-form pollen; and why this should be so it is difficult to conjecture, unless it be that the aboriginal parent-form of most heterostyled species possessed a pistil which exceeded its own stamens considerably in length.* I will only add that in a state of nature any single plant of a trimorphic species no doubt produces all three forms; and this may be accounted for either by its several flowers being separately fertilised by both the other forms, as Hildebrand supposes; or by pollen from both the other forms being deposited by insects on the stigma of the same flower.

Equal-styled varieties.—The tendency of the dimorphic species of Primula to produce equal-styled varieties deserves notice. Cases of this kind have

---

* It may be suspected that this was the case with Primula, judging from the length of the pistil in several allied genera (see Mr. J. Scott, 'Journal Linn. Soc. Bot.' vol. viii. 1864, p. 85). Herr Breitenbach found many specimens of *Primula elatior* growing in a state of nature with some flowers on the same plant long-styled, others short-styled and others equal-styled; and the long-styled form greatly preponderated in number; there being 61 of this form to 9 of the short-styled and 15 of the equal-styled.

been observed, as shown in the last chapter, in no less than six species, namely, *P. veris, vulgaris, Sinensis, auricula, farinosa,* and *elatior.* In the case of *P. veris,* the stamens resemble in length, position and size of their pollen-grains the stamens of the short-styled form; whilst the pistil closely resembles that of the long-styled, but as it varies much in length, one proper to the short-styled form appears to have been elongated and to have assumed at the same time the functions of a long-styled pistil. Consequently the flowers are capable of spontaneous self-fertilisation of a legitimate nature and yield a full complement of seed, or even more than the number produced by ordinary flowers legitimately fertilised. With *P. Sinensis,* on the other hand, the stamens resemble in all respects the shorter ones proper to the long-styled form, whilst the pistil makes a near approach to that of the short-styled, but as it varies in length, it would appear as if a long-styled pistil had been reduced in length and modified in function. The flowers in this case as in the last are capable of spontaneous legitimate fertilisation, and are rather more productive than ordinary flowers legitimately fertilised. With *P. auricula* and *farinosa* the stamens resemble those of the short-styled form in length, but those of the long-styled in the size of their pollen-grains; the pistil also resembles that of the long-styled, so that although the stamens and pistil are of nearly equal length, and consequently pollen is spontaneously deposited on the stigma, yet the flowers are not legitimately fertilised and yield only a very moderate supply of seed. We thus see, firstly, that equal-styled varieties have originated in various ways, and, secondly, that the combination of the two forms in the same flower differs in completeness. With *P. elatior* some of the flowers on the same plant have become equal-styled, instead of all of them as in the other species.

Mr. Scott has suggested that the equal-styled varieties arise through reversion to the former homostyled condition of the genus. This view is supported by the remarkable fidelity with which the equal-styled variation is transmitted after it has once appeared. I have shown in Chapter XIII. of my 'Variation of Animals and Plants under Domestication,' that any cause which disturbs the constitution tends to induce reversion, and it is chiefly the cultivated species of Primula which become equal-styled. Illegitimate fertilisation, which is an abnormal process, is likewise an exciting cause; and with illegitimately descended long-styled plants of *P. Sinensis*, I have observed the first appearance and subsequent stages of this variation. With some other plants of *P. Sinensis* of similar parentage the flowers appeared to have reverted to their original wild condition. Again, some hybrids between *P. veris* and *vulgaris* were strictly equal-styled, and others made a near approach to this structure. All these facts support the view that this variation results, at least in part, from reversion to the original state of the genus, before the species had become heterostyled. On the other hand, some considerations indicate, as previously remarked, that the aboriginal parent-form of Primula had a pistil which exceeded the stamens in length. The fertility of the equal-styled varieties has been somewhat modified, being sometimes greater and sometimes less than that of a legitimate union. Another view, however, may be taken with respect to the origin of the equal-styled varieties, and their appearance may be compared with that of hermaphrodites amongst

animals which properly have their sexes separated;
for the two sexes are combined in a monstrous hermaphrodite in a somewhat similar manner as the
two sexual forms are combined in the same flower of
an equal-styled variety of a heterostyled species.

Final remarks.—The existence of plants which have
been rendered heterostyled is a highly remarkable
phenomenon, as the two or three forms of the same
undoubted species differ not only in important points
of structure, but in the nature of their reproductive
powers. As far as structure is concerned, the two
sexes of many animals and of some plants differ to an
extreme degree; and in both kingdoms the same
species may consist of males, females, and hermaphrodites. Certain hermaphrodite cirripedes are aided in
their reproduction by a whole cluster of what I have
called complemental males, which differ wonderfully
from the ordinary hermaphrodite form. With ants
we have males and females, and two or three castes of
sterile females or workers. With Termites there are,
as Fritz Müller has shown, both winged and wingless
males and females, besides the workers. But in none
of these cases is there any reason to believe that the
several males or several females of the same species
differ in their sexual powers, except in the atrophied
condition of the reproductive organs in the workers of
social insects. Many hermaphrodite animals must
unite for reproduction, but the necessity of such
union apparently depends solely on their structure.
On the other hand, with heterostyled dimorphic
species there are two females and two sets of males,
and with trimorphic species three females and three
sets of males, which differ essentially in their sexual

powers. We shall, perhaps, best perceive the complex
and extraordinary nature of the marriage arrangements
of a trimorphic plant by the following illustration.
Let us suppose that the individuals of the same species
of ant always lived in triple communities; and that
in one of these, a large-sized female (differing also in
other characters) lived with six middle-sized and six
small-sized males ; in the second community a middle-
sized female lived with six large- and six small-sized
males ; and in the third, a small-sized female lived
with six large- and six middle-sized males. Each of
these three females, though enabled to unite with any
male, would be nearly sterile with her own two sets of
males, and likewise with two other sets of males of the
same size with her own which lived in the other two
communities ; but she would be fully fertile when
paired with a male of her own size. Hence the thirty-
six males, distributed by half-dozens in the three communities, would be divided into three sets of a dozen
each ; and these sets, as well as the three females,
would differ from one another in their reproductive
powers in exactly the same manner as do the distinct
species of the same genus. But it is a still more
remarkable fact that young ants raised from any one
of the three female ants, illegitimately fertilised by a
male of a different size would resemble in a whole
series of relations the hybrid offspring from a cross
between two distinct species of ants. They would be
dwarfed in stature, and more or less, or even utterly
barren. Naturalists are so much accustomed to behold
great diversities of structure associated with the two
sexes, that they feel no surprise at almost any amount
of difference ; but differences in sexual nature have
been thought to be the very touchstone of specific
distinction. We now see that such sexual differences

—the greater or less power of fertilising and being fertilised—may characterise the co-existing individuals of the same species, in the same manner as they characterise and have kept separate those groups of individuals, produced during the lapse of ages, which we rank and denominate as distinct species.

CHAPTER VII.

Polygamous, Diœcious, and Gyno-diœcious Plants.

The conversion in various ways of hermaphrodite into diœcious plants
—Heterostyled plants rendered diœcious—Rubiaceæ—Verbenaceæ
—Polygamous and sub-diœcious plants—Euonymus—Fragaria—
The two sub-forms of both sexes of Rhamnus and Epigæa—Ilex—
Gyno-diœcious plants—Thymus, difference in fertility of the hermaphrodite and female individuals—Satureia—Manner in which
the two forms probably originated—Scabiosa and other gyno-
diœcious plants—Difference in the size of the corolla in the forms
of polygamous, diœcious, and gyno-diœcious plants.

THERE are several groups of plants in which all the
species are diœcious, and these exhibit no rudiments
in the one sex of the organs proper to the other.
About the origin of such plants nothing is known. It
is possible that they may be descended from ancient
lowly organised forms, which had from the first their
sexes separated ; so that they have never existed as
hermaphrodites. There are, however, many other
groups of species and single ones, which from being
allied on all sides to hermaphrodites, and from ex-
hibiting in the female flowers plain rudiments of
male organs, and conversely in the male flowers rudiments of female organs, we may feel sure are descended
from plants which formerly had the two sexes combined in the same flower. It is a curious and obscure
problem how and why such hermaphrodites have been
rendered bisexual.

If in some individuals of a species the stamens
alone were to abort, females and hermaphrodites would

be left existing, of which many instances occur; and if the female organs of the hermaphrodite were afterwards to abort, the result would be a diœcious plant. Conversely, if we imagine the female organs alone to abort in some individuals, males and hermaphrodites would be left; and the hermaphrodites might afterwards be converted into females.

In other cases, as in that of the common Ash-tree mentioned in the Introduction, the stamens are rudimentary in some individuals, the pistils in others, others again remaining as hermaphrodites. Here the modification of the two sets of organs appears to have occurred simultaneously, as far as we can judge from their equal state of abortion. If the hermaphrodites were supplanted by the individuals having separated sexes, and if these latter were equalised in number, a strictly diœcious species would be formed.

There is much difficulty in understanding why hermaphrodite plants should ever have been rendered diœcious. There would be no such conversion, unless pollen was already carried regularly by insects or by the wind from one individual to the other; for otherwise every step towards diœciousness would lead towards sterility. As we must assume that cross-fertilisation was assured before an hermaphrodite could be changed into a diœcious plant, we may conclude that the conversion has not been effected for the sake of gaining the great benefits which follow from cross-fertilisation. We can, however, see that if a species were subjected to unfavourable conditions from severe competition with other plants, or from any other cause, the production of the male and female elements and the maturation of the ovules by the same individual, might prove too great a strain on its powers, and the separation of the sexes would then be highly beneficial.

This, however, would be effected only under the con-
tingency of a reduced number of seeds, produced by
the females alone, being sufficient to keep up the
stock.

There is another way of looking at the subject which
partially removes a difficulty that appears at first sight
insuperable, namely, that during the conversion of an
hermaphrodite into a diœcious plant, the male organs
must abort in some individuals and the female organs
in others. Yet as all are exposed to the same conditions, it might have been expected that those
which varied would tend to vary in the same manner. As a general rule only a few individuals of a
species vary simultaneously in the same manner; and
there is no improbability in the assumption that
some few individuals might produce larger seeds
than the average, better stocked with nourishment. If
the production of such seeds were highly beneficial to
a species, and on this head there can be little doubt,*
the variety with the large seeds would tend to increase. But in accordance with the law of compensation we might expect that the individuals which produced such seeds would, if living under severe conditions, tend to produce less and less pollen, so that
their anthers would be reduced in size and might ultimately become rudimentary. This view occurred to
me owing to a statement by Sir J. E. Smith † that
there are female and hermaphrodite plants of *Serratula
tinctoria*, and that the seeds of the former are larger
than those of the hermaphrodite form. It may also
be worth while to recall the case of the mid-styled
form of *Lythrum salicaria*, which produces a larger

---

* See the facts given in 'The
Effects of Cross and Self-fertilisation,' p. 353.

† 'Trans. Linn. Soc.,' vol. xiii.
p. 600.

number of seeds than the other forms, and has somewhat smaller pollen-grains which have less fertilising
power than those of the corresponding stamens in the
other two forms; but whether the larger number of
seeds is the indirect cause of the diminished power
of the pollen, or *vice versâ*, I know not. As soon
as the anthers in a certain number of individuals became reduced in size in the manner just suggested or
from any other cause, the other individuals would have
to produce a larger supply of pollen; and such increased development would tend to reduce the female
organs through the law of compensation, so as ultimately to leave them in a rudimentary condition;
and the species would then become diœcious.

Instead of the first change occurring in the female
organs we may suppose that the male ones first varied,
so that some individuals produced a larger supply of
pollen. This would be beneficial under certain circumstances, such as a change in the nature of the
insects which visited the flowers, or in their becoming more anemophilous, for such plants require an
enormous quantity of pollen. The increased action of
the male organs would tend to affect through compensation the female organs of the same flower; and the
final result would be that the species would consist of
males and hermaphrodites. But it is of no use considering this case and other analogous ones, for, as
stated in the Introduction, the co-existence of male
and hermaphrodite plants is excessively rare.

It is no valid objection to the foregoing views that
changes of such a nature would be effected with extreme slowness, for we shall presently see good reason
to believe that various hermaphrodite plants have
become or are becoming diœcious by many and excessively small steps. In the case of polygamous

species, which exist as males, females and hermaphrodites, the latter would have to be supplanted before the species could become strictly diœcious; but the extinction of the hermaphrodite form would probably not be difficult, as a complete separation of the sexes appears often to be in some way beneficial. The males and females would also have to be equalised in number, or produced in some fitting proportion for the effectual fertilisation of the females.

There are, no doubt, many unknown laws which govern the suppression of the male or female organs in hermaphrodite plants, quite independently of any tendency in them to become monœcious, diœcious, or polygamous. We see this in those hermaphrodites which from the rudiments still present manifestly once possessed more stamens or pistils than they now do,—even twice as many, as a whole verticil has often been suppressed. Robert Brown remarks* that "the order of reduction or abortion of the stamina in any natural family may with some confidence be predicted," by observing in other members of the family, in which their number is complete, the order of the dehiscence of the anthers; for the lesser permanence of an organ is generally connected with its lesser perfection, and he judges of perfection by priority of development. He also states that whenever there is a separation of the sexes in an hermaphrodite plant, which bears flowers on a simple spike, it is the females which expand first; and this he likewise attributes to the female sex being the more perfect of the two, but why the female should be thus valued he does not explain.

* 'Trans. Linn. Soc.' vol. xii. p. 98.   Or 'Miscellaneous Works,' vol. ii. pp. 278–81.

Plants under cultivation or changed conditions of life frequently become sterile; and the male organs are much oftener affected than the female, though the latter alone are sometimes affected. The sterility of the stamens is generally accompanied by a reduction in their size; and we may feel sure, from a wide-spread analogy, that both the male and female organs would become rudimentary in the course of many generations if they failed altogether to perform their proper functions. According to Gärtner,* if the anthers on a plant are contabescent (and when this occurs it is always at a very early period of growth) the female organs are sometimes precociously developed. I mention this case as it appears to be one of compensation. So again is the well-known fact, that plants which increase largely by stolons or other such means are often utterly barren, with a large proportion of their pollen-grains in a worthless condition.

Hildebrand has shown that with hermaphrodite plants which are strongly proterandrous, the stamens in the flowers which open first sometimes abort; and this seems to follow from their being useless, as no pistils are then ready to be fertilised. Conversely the pistils in the flowers which open last sometimes abort; as when they are ready for fertilisation all the pollen has been shed. He further shows by means of a series of gradations amongst the Compositæ,† that a tendency from the causes just specified to produce either male or female florets, sometimes spreads to all the florets on the same head, and sometimes

* 'Beiträge zur Kenntniss,' &c. p. 117 et seq. The whole subject of the sterility of plants from various causes has been discussed in my 'Variation of Animals and Plants under Domestication,' chap. xviii.—2nd edit. vol. ii. pp. 146–56.

† 'Ueber die Geschlechtsverhältnisse bei den Compositen,' 1869, p. 89.

even to the whole plant; and in this latter case the
species becomes diœcious. In those rare instances mentioned in the Introduction, in which some of the individuals of both monœcious and hermaphrodite plants
are proterandrous, others being proterogynous, their
conversion into a diœcious condition would probably be
much facilitated, as they already consist of two bodies
of individuals, differing to a certain extent in their
reproductive functions.

Dimorphic heterostyled plants offer still more
strongly marked facilities for becoming diœcious; for
they likewise consist of two bodies of individuals in
approximately equal numbers, and what probably is
more important, both the male and female organs
differ in the two forms, not only in structure but in
function, in nearly the same manner as do the reproductive organs of two distinct species belonging to
the same genus. Now if two species are subjected to
changed conditions, though of the same nature, it is
notorious that they are often affected very differently;
therefore the male organs, for instance, in one form of
a heterostyled plant might be affected by those unknown causes which induce abortion, differently from
the homologous but functionally different organs in
the other form; and so conversely with the female
organs. Thus the great difficulty before alluded to is
much lessened in understanding how any cause whatever could lead to the simultaneous reduction and
ultimate suppression of the male organs in half the
individuals of a species, and of the female organs in
the other half, whilst all were subjected to exactly the
same conditions of life.

That such reduction or suppression has occurred
in some heterostyled plants is almost certain. The
Rubiaceæ contain more heterostyled genera than any

other family, and from their wide distribution we may
infer that many of them became heterostyled at a remote period, so that there will have been ample time for
some of the species to have been since rendered diœcious. Asa Gray informs me that Coprosma is diœcious,
and that it is closely allied through Nertera to Mitchella, which as we know is a heterostyled dimorphic
species. In the male flowers of Coprosma the stamens
are exserted, and in the female flowers the stigmas;
so that, judging from the affinities of the above three
genera, it seems probable that an ancient short-styled
form bearing long stamens with large anthers and
large pollen-grains (as in the case of several Rubiaceous genera) has been converted into the male Coprosma; and that an ancient long-styled form with short
stamens, small anthers and small pollen-grains has
been converted into the female form. But according
to Mr. Meehan,* Mitchella itself is diœcious in some
districts; for he says that one form has small sessile
anthers without a trace of pollen, the pistil being
perfect; while in another form the stamens are perfect
and the pistil rudimentary. He adds that plants
may be observed in the autumn bearing an abundant
crop of berries, and others without a single one.
Should these statements be confirmed, Mitchella will
be proved to be heterostyled in one district and
diœcious in another.

Asperula is likewise a Rubiaceous genus, and from
the published description of the two forms of *A. sco-
paria,* an inhabitant of Tasmania, I did not doubt that
it was heterostyled; but on examining some flowers
sent me by Dr. Hooker they proved to be diœcious.
The male flowers have large anthers and a very small

* ' Proc. Acad. of Sciences of Philadelphia,' July 28, 1868, p. 183.

ovarium, surmounted by a mere vestige of a stigma without any style; whilst the female flowers possess a large ovarium, the anthers being rudimentary and apparently quite destitute of pollen. Considering how many Rubiaceous genera are heterostyled, it is a reasonable suspicion that this Asperula is descended from a heterostyled progenitor; but we should be cautious on this head, for there is no improbability in a homostyled Rubiaceous plant becoming diœcious. Moreover, in an allied plant, *Galium cruciatum*, the female organs have been suppressed in most of the lower flowers, whilst the upper ones remain hermaphrodite; and here we have a modification of the sexual organs without any connection with heterostylism.

Mr. Thwaites informs me that in Ceylon various Rubiaceous plants are heterostyled; but in the case of Discospermum one of the two forms is always barren, the ovary containing about two aborted ovules in each loculus; whilst in the other form each loculus contains several perfect ovules; so that the species appears to be strictly diœcious.

Most of the species of the South American genus Ægiphila, a member of the Verbenaceæ, apparently are heterostyled; and both Fritz Müller and myself thought that this was the case with *Æ. obdurata*, so closely did its flowers resemble those of the heterostyled species. But on examining the flowers, the anthers of the long-styled form were found to be entirely destitute of pollen and less than half the size of those in the other form, the pistil being perfectly developed. On the other hand, in the short-styled form the stigmas are reduced to half their proper length, having also an abnormal appearance; whilst the stamens are perfect. This plant therefore is diœcious; and we may, I think, conclude that a short-styled progenitor, bearing long stamens exserted beyond the corolla, has been converted into the male; and a long-styled progenitor with fully developed stigmas into the female.

From the number of bad pollen-grains in the small anthers of the short stamens of the long-styled form of *Pulmonaria angustifolia*, we may suspect that this form is tending to become female; but it does not appear that the other or short-styled form is becoming more masculine. Certain appearances countenance the belief that the reproductive system of *Phlox subulata* is likewise undergoing a change of some kind.

I have now given the few cases known to me in which heterostyled plants appear with some considerable degree of probability to have been rendered diœcious. Nor ought we to expect to find many such cases, for the number of heterostyled species is by no means large, at least in Europe, where they could hardly have escaped notice. Therefore the number of diœcious species which owe their origin to the transformation of heterostyled plants is probably not so large as might have been anticipated from the facilities which they offer for such conversion.

In searching for cases like the foregoing ones, I have been led to examine some diœcious or sub-diœcious plants, which are worth describing, chiefly as they show by what fine gradations hermaphrodites may pass into polygamous or diœcious species.

Polygamous, Diœcious and Sub-diœcious Plants.

Euonymus Europæus (Celastrineæ).—The spindle-tree is described in all the botanical works which I have consulted as an hermaphrodite. Asa Gray speaks of the flowers of the American species as perfect, whilst those in the allied genus Celastrus are said to be "polygamo-diœcious." If a number of bushes of our spindle-tree be examined, about half will be found to have stamens equal in length to the pistil, with well-developed anthers; the pistil being likewise to all appearance well developed. The other half have a perfect pistil, with the stamens short, bearing rudimentary anthers destitute of pollen; so that these bushes are females. All the flowers on the same plant present the same structure. The female corolla is smaller than that on the polleniferous bushes. The two forms are shown in the accompanying drawings.

Fig. 12.

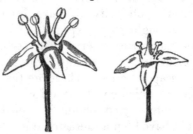

Hermaphrodite or male.   Female.
EUONYMUS EUROPÆUS.

I did not at first doubt that this species existed under an hermaphrodite and female form; but we shall presently see that some of the bushes which appear to be hermaphrodites never produce fruit, and these are in fact males. The species, therefore, is polygamous in the sense in which I use the term, and trioicous. The flowers are frequented by many Diptera and some small Hymenoptera for the sake of the nectar secreted by the disc, but I did not see a single bee at work; nevertheless the other insects sufficed to fertilise effectually female bushes growing at a distance of even 30 yards from any polleniferous bush.

The small anthers borne by the short stamens of the female flowers are well formed and dehisce properly, but I could never find in them a single grain of pollen. It is somewhat difficult to compare the length of the pistils in the two forms, as they vary somewhat in this respect and continue to grow after the anthers are mature. The pistils, therefore, in old flowers on a polleniferous plant are often of considerably greater length than in young flowers on a female plant. On this account the pistils from five flowers from so many hermaphrodite or male bushes were compared with those from five female bushes, before the anthers had dehisced and whilst the rudimentary ones were of a pink colour and not at all shrivelled. These two sets of pistils did not differ in length, or if there was any difference those of the polleniferous flowers were rather the longest. In one hermaphrodite plant, which produced during three years very few and poor fruit, the pistil much exceeded in length the stamens bearing perfect and as yet closed anthers; and I never saw such a case on any female plant. It is a surprising fact that the pistil in the male and in the semi-sterile hermaphrodite flowers has not been reduced in length, seeing that it performs very poorly or not at all its proper function. The stigmas in the two forms are exactly alike; and in some of the polleniferous plants which never produced any fruit I found that the surface of the stigma was viscid, so that pollen-grains adhered to it and had exserted their tubes. The ovules are of equal size in the two forms. Therefore the most acute botanist, judging only by structure, would never have suspected

U

that some of the bushes were in function exclusively males.

Thirteen bushes growing near one another in a hedge consisted of eight females quite destitute of pollen and of five hermaphrodites with well-developed anthers. In the autumn the eight females were well covered with fruit, excepting one, which bore only a moderate number. Of the five hermaphrodites, one bore a dozen or two fruits, and the remaining four bushes several dozen; but their number was as nothing compared with those on the female bushes, for a single branch, between two and three feet in length, from one of the latter, yielded more than any one of the hermaphrodite bushes. The difference in the amount of fruit produced by the two sets of bushes is all the more striking, as from the sketches above given it is obvious that the stigmas of the polleniferous flowers can hardly fail to receive their own pollen; whilst the fertilisation of the female flowers depends on pollen being brought to them by flies and the smaller Hymenoptera, which are far from being such efficient carriers as bees.

I now determined to observe more carefully during successive seasons some bushes growing in another place about a mile distant. As the female bushes were so highly productive, I marked only two of them with the letters A and B, and five polleniferous bushes with the letters C to G. I may premise that the year 1865 was highly favourable for the fruiting of all the bushes, especially for the polleniferous ones, some of which were quite barren except under such favourable conditions. The season of 1864 was unfavourable. In 1863 the female A produced "some fruit;" in 1864 only 9; and in 1865, 97 fruit. The female B in 1863 was "covered with fruit;" in 1864 it bore 28; and in

1865 "innumerable very fine. fruits." I may add, that three other female trees growing close by were observed, but only during 1863, and they then bore abundantly. With respect to the polleniferous bushes, the one marked C did not bear a single fruit during the years 1863 and 1864, but during 1865 it produced no less than 92 fruit, which, however, were very poor. I selected one of the finest branches with 15 fruit, and these contained 20 seeds, or on an average 1·33 per fruit. I then took by hazard 15 fruit from an adjoining female bush, and these contained 43 seeds; that is, more than twice as many, or on an average 2·86 per fruit. Many of the fruits from the female bushes included four seeds, and only one had a single seed; whereas not one fruit from the polleniferous bushes contained four seeds. Moreover when the two lots of seeds were compared, it was manifest that those from the female bushes were the larger. The second polleniferous bush, D, bore in 1863 about two dozen fruit,—in 1864 only 3 very poor fruit, each containing a single seed,—and in 1865, 20 equally poor fruit. Lastly, the three polleniferous bushes, E, F, and G, did not produce a single fruit during the three years 1863, 1864, and 1865.

We thus see that the female bushes differ somewhat in their degree of fertility, and the polleniferous ones in the most marked manner. We have a perfect gradation from the female bush, B, which in 1865 was covered with " innumerable fruits," — through the female A, which produced during the same year 97,— through the polleniferous bush C, which produced this year 92 fruits, these, however, containing a very low average number of seeds of small size,—through the bush D, which produced only 20 poor fruit,—to the three bushes, E, F, and G, which did not this

year, or during the two previous years, produce a
single fruit. If these latter bushes and the more
fertile female ones were to supplant the others, the
spindle-tree would be as strictly diœcious in function
as any plant in the world. This case appears to me
very interesting, as showing how gradually an hermaphrodite plant may be converted into a diœcious one.*

Seeing how general it is for organs which are
almost or quite functionless to be reduced in size, it is
remarkable that the pistils of the polleniferous plants
should equal or even exceed in length those of the
highly fertile female plants. This fact formerly led
me to suppose that the spindle-tree had once been
heterostyled; the hermaphrodite and male plants having been originally long-styled, with the pistils since
reduced in length, but with the stamens retaining
their former dimensions; whilst the female plant had
been originally short-styled, with the pistil in its present state, but with the stamens since greatly reduced
and rendered rudimentary. A conversion of this kind
is at least possible, although it is the reverse of
that which appears actually to have occurred with
some Rubiaceous genera and Ægiphila; for with these
plants the short-styled form has become the male, and
the long-styled the female. It is, however, a more
simple view that sufficient time has not elapsed for the

---

* According to Fritz Müller
('Bot. Zeitung,' 1870, p. 151), a
Chamissoa (Amaranthaceæ) in
Southern Brazil is in nearly the
same state as our Euonymus. The
ovules are equally developed in the
two forms. In the female the pistil
is perfect, whilst the anthers are
entirely destitute of pollen. In
the polleniferous form, the pistil
is short and the stigmas never
separate from one another, so
that, although their surfaces are
covered with fairly well-developed
papillæ, they cannot be fertilised.
These latter plants do not commonly yield any fruit, and are
therefore in function males. Nevertheless, on one occasion Fritz
Müller found flowers of this kind in
which the stigmas had separated,
and they produced some fruit.

reduction of the pistil in the male and hermaphrodite flowers of our Euonymus; though this view does not account for the pistils in the polleniferous flowers being sometimes longer than those in the female flowers.

Fragaria vesca, Virginiana, Chiloensis, &c. (Rosaceæ). —A tendency to the separation of the sexes in the cultivated strawberry seems to be much more strongly marked in the United States than in Europe; and this appears to be the result of the direct action of climate on the reproductive organs. In the best ac- count which I have seen,[*] it is stated that many of the varieties in the United States consist of three forms, namely, females, which produce a heavy crop of fruit,— of hermaphrodites, which " seldom produce other than a very scanty crop of inferior and imperfect berries," —and of males, which produce none. The most skilful cultivators plant " seven rows of female plants, then one row of hermaphrodites, and so on throughout the field." The males bear large, the hermaphrodites mid-sized, and the females small flowers. The latter plants produce few runners, whilst the two other forms produce many; consequently, as has been observed both in England and in the United States, the polleni- ferous forms increase rapidly and tend to supplant the females. We may therefore infer that much more vital force is expended in the production of ovules and fruit than in the production of pollen. Another species, the Hautbois strawberry (*F. elatior*), is more strictly diœcious; but Lindley made by selection an hermaphrodite stock.[†]

Rhamnus catharticus (Rhamneæ).—This plant is well

---

[*] Mr. Leonard Wray in ' Gard. Chron.' 1861, p. 716.

[†] For references and further information on this subject, see ' Variation under Domestication,' chap. x. 2nd edit. vol. i. p. 375.

known to be diœcious. My son William found the
two sexes growing in about equal numbers in the Isle
of Wight, and sent me specimens, together with observations on them. Each sex consists of two sub-forms.
The two forms of the male differ in their pistils:
in some plants it is quite small, without any distinct
stigma; in others the pistil is much more developed,
with the papillæ on the stigmatic surfaces moderately
large. The ovules in both kinds of males are in an
aborted condition. On my mentioning this case to Professor Caspary, he examined several male plants in
the botanic gardens at Königsberg, where there were
no females, and sent me the accompanying drawings.

Fig. 13.

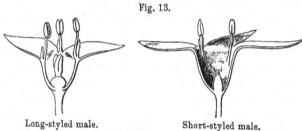

Long-styled male.   Short-styled male.
RHAMNUS CATHARTICUS. (From Caspary.)

In the English plants the petals are not so greatly
reduced as represented in this drawing. My son observed that those males which had their pistils moderately well developed bore sliglily larger flowers, and,
what is very remarkable, their pollen-grains exceeded
by a little in diameter those of the males with greatly
reduced pistils. This fact is opposed to the belief that
the present species was once heterostyled; for in this
case it might have been expected that the shorter-
styled plants would have had larger pollen-grains.

In the female plants the stamens are in an extremely rudimentary condition, much more so than

the pistils in the males.   The pistil varies considerably in length in the female plants, so that they
may be divided into two sub-forms according to the

Fig. 14.

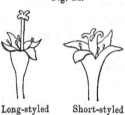

Long-styled   Short-styled
female.   female.

RHAMNUS CATHARTICUS.

length of this organ.   Both the petals and sepals are
decidedly smaller in the females than in the males ;
and the sepals do not turn downwards, as do those of
the male flowers when mature.   All the flowers on the
same male or same female bush, though subject to
some variability, belong to the same sub-form ; and
as my son never experienced any difficulty in decid-
ing under which class a plant ought to be included,
he believes that the two sub-forms of the same sex
do not graduate into one another.   I can form no
satisfactory theory how the four forms of this plant
originated.

Rhamnus lanceolatus exists in the United States,
as I am informed by Professor Asa Gray, under two
hermaphrodite forms.   In the one, which may be called
the short-styled, the flowers are sub-solitary, and include a pistil about two-thirds or only half as long as
that in the other form ; it has also shorter stigmas.   The
stamens are of equal length in the two forms ; but the
anthers of the short-styled contain rather less pollen,
as far as I could judge from a few dried flowers.   My

son compared the pollen-grains from the two forms,
and those from the long-styled flowers were to those
from the short-styled, on an average from ten measurements, as 10 to 9 in diameter; so that the two hermaphrodite forms of this species resemble in this
respect the two male forms of *R. catharticus.* The
long-styled form is not so common as the short-styled.
The latter is said by Asa Gray to be the more fruitful
of the two, as might have been expected from its
appearing to produce less pollen, and from the grains
being of smaller size; it is therefore the more highly
feminine of the two. The long-styled form produces
a greater number of flowers, which are clustered together instead of being sub-solitary; they yield some
fruit, but as just stated are less fruitful than the other
form, so that this form appears to be the more masculine of the two. On the supposition that we have
here an hermaphrodite plant becoming diœcious, there
are two points deserving notice; firstly, the greater
length of the pistil in the incipient male form; and
we have met with a nearly similar case in the male
and hermaphrodite forms of Euonymus compared with
the females. Secondly, the larger size of the pollen-
grains in the more masculine flowers, which perhaps may
be attributed to their having retained their normal size;
whilst those in the incipient female flowers have been
reduced. The long-styled form of *R. lanceolatus* seems
to correspond with the males of *R. catharticus* which
have a longer pistil and larger pollen-grains. Light
will perhaps be thrown on the nature of the forms
in this genus, as soon as the power of both kinds of
pollen on both stigmas is ascertained. Several other
species of Rhamnus are said to be diœcious* or sub-

---

* Lecoq, 'Géogr. Bot.' tom. v. 1856, pp. 420–26.

diœcious. On the other hand, *R. frangula* is an ordinary hermaphrodite, for my son found a large number of bushes all bearing an equal profusion of fruit.

Epigæa repens (Ericaceæ).—This plant appears to be in nearly the same state as *Rhamnus catharticus*. It is described by Asa Gray* as existing under four forms. (1) With long style, perfect stigma, and short abortive stamens. (2) Shorter style, but with stigma equally perfect, short abortive stamens. These two female forms amounted to 20 per cent. of the specimens received from one locality in Maine; but all the fruiting specimens belonged to the first form. (3) Style long, as in No. 1, but with stigma imperfect, stamens perfect. (4) Style shorter than in the last, stigma imperfect, stamens perfect. These two latter forms are evidently males. Therefore, as Asa Gray remarks, "the flowers may be classified into two kinds, each with two modifications; the two main kinds characterised by the nature and perfection of the stigma, along with more or less abortion of the stamens; their modifications, by the length of the style." Mr. Meehan has described† the extreme variability of the corolla and calyx in this plant, and shows that it is diœcious. It is much to be wished that the pollen-grains in the two male forms should be compared, and their fertilising power tried on the two female forms.

Ilex aquifolium (Aquifoliaceæ). — In the several works which I have consulted, one author alone‡ says that the holly is diœcious. During several years I

---

* 'American Journal of Science,' July 1876. Also 'The American Naturalist,' 1876, p. 490.
† "Variations in *Epigæa repens*," 'Proc. Acad. Nat. Soc. of Phila-delphia,' May 1868, p. 153.
‡ Vaucher, 'Hist. Phys. des Plantes d'Europe,' 1841, tom. ii. p. 11.

have examined many plants, but have never found one that was really hermaphrodite. I mention this genus because the stamens in the female flowers, although quite destitute of pollen, are but slightly and sometimes not at all shorter than the perfect stamens in the male flowers. In the latter the ovary is small and the pistil is almost aborted. The filaments of the perfect stamens adhere for a greater length to the petals than in the female flowers. The corolla of the latter is rather smaller than that of the male. The male trees produce a greater number of flowers than the females. Asa Gray informs me that *I. opaca*, which represents in the United States our common holly, appears (judging from dried flowers) to be in a similar state ; and so it is, according to Vaucher, with several other but not with all the species of the genus.

Gyno-diœcious Plants.

The plants hitherto described either show a tendency to become diœcious, or apparently have become so within a recent period. But the species now to be considered consist of hermaphrodites and females without males, and rarely show any tendency to be diœcious, as far as can be judged from their present condition and from the absence of species having separated sexes within the same groups. Species belonging to the present class, which I have called gyno-diœcious, are found in various widely distinct families; but are much more common in the Labiatæ (as has long been noticed by botanists) than in any other group. Such cases have been noticed by myself in *Thymus serpyllum* and *vulgaris, Satureia hortensis, Origanum vulgare,* and *Mentha hirsuta ;* and by others in *Nepeta glechoma, Mentha vulgaris* and

aquatica, and *Prunella vulgaris*. In these two latter
species the female form, according to H. Müller, is
infrequent. To these must be added *Dracocephalum
Moldavicum*, *Melissa officinalis* and *clinipodium*, and
Hyssopus officinalis.* In the two last-named plants the
female form likewise appears to be rare, for I raised
many seedlings of both, and all were hermaphrodites.
It has already been remarked in the Introduction that
andro-dioecious species, as they may be called, or those
which consist of hermaphrodites and males, are extremely rare, or hardly exist.

Thymus serpyllum.—The hermaphrodite plants present nothing particular in the state of their reproductive organs ; and so it is in all the following cases. The
females of the present species produce rather fewer
flowers and have somewhat smaller corollas than the
hermaphrodites ; so that near Torquay, where this
plant abounds, I could, after a little practice, distinguish the two forms whilst walking quickly past them.
According to Vaucher, the smaller size of the corolla
is common to the females of most or all of the above-
mentioned Labiatæ. The pistil of the female, though
somewhat variable in length, is generally shorter,
with the margins of the stigma broader and formed
of more lax tissue, than that of the hermaphrodite.
The stamens in the female vary excessively in length ;
they are generally enclosed within the tube of the

---

* H. Müller, 'Die Befruchtung
der Blumen,' 1873 ; and 'Nature,'
1873, p. 161. Vaucher, 'Plantes
d'Europe,' tom. iii. p. 611. For
Dracocephalum, Schimper, as
quoted by Braun, 'Annals and
Mag. of Nat. Hist.' 2nd series, vol.
xviii. 1856, p. 380. Lecoq, 'Géographie Bot. de l'Europe,' tom. viii.
pp. 33, 38, 44, &c. Both Vaucher
and Lecoq were mistaken in think-
ing that several of the plants
named in the text are dioecious.
They appear to have assumed that
the hermaphrodite form was a
male ; perhaps they were deceived by the pistil not becoming
fully developed and of proper
length until some time after the
anthers have dehisced.

corolla, and their anthers do not contain any sound pollen; but after long search I found a single plant with the stamens moderately exserted, and their anthers contained a very few full-sized grains, together with a multitude of minute empty ones. In some females the stamens are extremely short, and their minute anthers, though divided into the two normal cells or loculi, contained not a trace of pollen: in others again the anthers did not exceed in diameter the filaments which supported them, and were not divided into two loculi. Judging from what I have myself seen and from the descriptions of others, all the plants in Britain, Germany, and near Mentone, are in the state just described; and I have never found a single flower with an aborted pistil. It is, therefore, remarkable that, according to Delpino,* this plant near Florence is generally trimorphic, consisting of males with aborted pistils, females with aborted stamens, and hermaphrodites.

I found it very difficult to judge of the proportional number of the two forms at Torquay. They often grow mingled together, but with large patches consisting of one form alone. At first I thought that the two were nearly equal in number; but on examining every plant which grew close to the edge of a little overhanging dry cliff, about 200 yards in length, I found only 12 females; all the rest, some hundreds in number, being hermaphrodites. Again, on an extensive gently sloping bank, which was so thickly covered with this plant that, viewed from the distance of half a mile it appeared of a pink colour, I could not discover a single female. Therefore the her-

---

* 'Sull' Opera, la Distribuzione dei Sessi nelle Piante, &c.' 1867, p. 7. With respect to Germany, H. Müller, 'Die Befruchtung,' &c.,' p. 327.

maphrodites must greatly exceed in number the females, at least in the localities examined by me. A very dry station apparently favours the presence of the female form. With some of the other above-named Labiatæ the nature of the soil or climate likewise seems to determine the presence of one or both forms; thus with *Nepeta glechoma*, Mr. Hart found in 1873 that all the plants which he examined near Kilkenny in Ireland were females; whilst all near Bath were hermaphrodites, and near Hertford both forms were present, but with a preponderance of hermaphrodites.* It would, however, be a mistake to suppose that the nature of the conditions determines the form independently of inheritance; for I sowed in the same small bed seeds of *T. serpyllum*, gathered at Torquay from the female alone, and these produced an abundance of both forms. There is every reason to believe, from large patches consisting of the same form, that the same individual plant, however much it may spread, always retains the same form. In two distant gardens I found masses of the lemon-thyme (*T. citriodorus*, a var. of *T. serpyllum*), which I was informed had grown there during many years, and every flower was female.

With respect to the fertility of the two forms, I marked at Torquay a large hermaphrodite and a large female plant of nearly equal sizes, and when the seeds were ripe I gathered all the heads. The two heaps were of very nearly equal bulk; but the heads from the female plant numbered 160, and their seeds weighed 8·7 grains; whilst those from the hermaphrodite plant numbered 200, and their seeds weighed only 4·9 grains; so that the seeds from the

---

* 'Nature,' June 1873, p. 162.

female plant were to those from the hermaphrodite
as 100 to 56 in weight. If the relative weight of
the seeds from an equal number of flower-heads from
the two forms be compared, the ratio is as 100 for the
female to 45 for the hermaphrodite form.

Thymus vulgaris.—The common garden thyme resembles in almost every respect *T. serpyllum.* The
same slight differences between the stigmas of the
two forms could be perceived. In the females the
stamens are not generally quite so much reduced as
in the same form of *T. serpyllum.* In some specimens
sent me from Mentone by Mr. Moggridge, together
with the accompanying sketches, the anthers of the

Fig. 15.

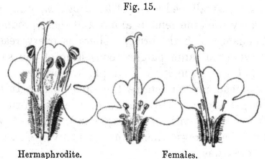

Hermaphrodite.   Females.
THYMUS VULGARIS (magnified).

female, though small, were well formed, but they contained very little pollen, and not a single sound grain
could be detected. Eighteen seedlings were raised
from purchased seed, sown in the same small bed;
and these consisted of seven hermaphrodites and
eleven females. They were left freely exposed to
the visits of bees, and no doubt every female flower
was fertilised; for on placing under the microscope a large number of stigmas from female plants,

not one could be found to which pollen-grains of thyme did not adhere. The seeds were carefully collected from the eleven female plants, and they weighed 98·7 grains; and those from the seven hermaphrodites 36·5 grains. This gives for an equal number of plants the ratio of 100 to 58; and we here see, as in the last case, how much more fertile the females are than the hermaphrodites. These two lots of seeds were sown separately in two adjoining beds, and the seedlings from both the hermaphrodite and female parent-plants consisted of both forms.

Satureia hortensis.—Eleven seedlings were raised in separate pots in a hotbed and afterwards kept in the green-house. They consisted of ten females and of a single hermaphrodite. Whether or not the conditions to which they had been subjected caused the great excess of females I do not know. In the females the pistil is rather longer than that of the hermaphrodite, and the stamens are mere rudiments, with minute colourless anthers destitute of pollen. The windows of the green-house were left open, and the flowers were incessantly visited by humble and hive bees. Although the ten females did not produce a single grain of pollen, yet they were all thoroughly well fertilised by the one hermaphrodite plant, and this is an interesting fact. It should be added that no other plant of this species grew in my garden. The seeds were collected from the finest female plant, and they weighed 78 grains; whilst those from the hermaphrodite, which was a rather larger plant than the female, weighed only 33·2 grains; that is, in the ratio of 100 to 43. The female form, therefore, is very much more fertile than the hermaphrodite, as in the two last cases; but the hermaphrodite was necessarily self-fertilised, and this probably diminished its fertility.

We may now consider the probable means by which
so many of the Labiatæ have been separated into two
forms, and the advantages thus gained. H. Müller*
supposes that originally some individuals varied so as
to produce more conspicuous flowers; and that insects
habitually visited these first, and then dusted with
their pollen visited and fertilised the less conspicuous
flowers. The production of pollen by the latter plants
would thus be rendered superfluous, and it would be
advantageous to the species that their stamens should
abort, so as to save useless expenditure. They would
thus be converted into females. But another view may
be suggested: as the production of a large supply of
seeds evidently is of high importance to many plants,
and as we have seen in the three foregoing cases
that the females produce many more seeds than the
hermaphrodites, increased fertility seems to me the
more probable cause of the formation and separation of
the two forms. From the data above given it follows
that ten plants of *Thymus serpyllum*, if half consisted
of hermaphrodites and half of females, would yield
seeds compared with ten hermaphrodite plants in the
ratio of 100 to 72. Under similar circumstances the
ratio with *Satureia hortensis* (subject to the doubt from
the self-fertilisation of the hermaphrodite) would be as
100 to 60. Whether the two forms originated in certain individuals varying and producing more seed than
usual, and consequently producing less pollen; or in
the stamens of certain individuals tending from some
unknown cause to abort, and consequently producing
more seed, it is impossible to decide; but in either
case, if the tendency to the increased production of
seed were steadily favoured, the result would be the

* 'Die Befruchtung der Blumen,' pp. 319, 326.

complete abortion of the male organs. I shall presently discuss the cause of the smaller size of the
female corolla.

Scabiosa arvensis (Dipsaceæ).—It has been shown by H. Müller
that this species exists in Germany under an hermaphrodite and
female form.* In my neighbourhood (Kent) the female plants
do not nearly equal in number the hermaphrodites. The stamens
of the females vary much in their degree of abortion; in some
plants they are quite short and produce no pollen; in others
they reach to the mouth of the corolla, but their anthers are not
half the proper size, never dehisce, and contain but few pollen-
grains, these being colourless and of small diameter. The hermaphrodite flowers are strongly proterandous, and H. Müller
shows that, whilst all the stigmas on the same flower-head are
mature at nearly the same time, the stamens dehisce one after
the other; so that there is a great excess of pollen, which serves
to fertilise the female plants. As the production of pollen by
one set of plants is thus rendered superfluous, their male organs
have become more or less completely aborted. Should it be
hereafter proved that the female plants yield, as is probable,
more seeds than the hermaphrodites, I should be inclined to
extend the same view to this plant as to the Labiatæ. I have
also observed the existence of two forms in our endemic *S.
succisa*, and in the exotic *S. atro-purpurea*. In the latter plant,
differently to what occurs in *S. arvensis*, the female flowers,
especially the larger circumferential ones, are smaller than those
of the hermaphrodite form. According to Lecoq, the female
flower-heads of *S. succisa* are likewise smaller than those of
what he calls the male plants, but which are probably hermaphrodites.

Echium vulgare (Boragineæ).—The ordinary hermaphrodite
form appears to be proterandrous, and nothing more need be said
about it. The female differs in having a much smaller corolla
and shorter pistil, but a well-developed stigma. The stamens

---

* 'Befruchtung der Blumen,'
&c., p. 368. The two forms occur
not only in Germany, but in
England and France. Lecoq,
('Géographie Bot.' 1857, tom. vi.
pp. 473, 477) says that male
plants as well as hermaphrodites
and females co-exist; it is, however, possible that he may have
been deceived by the flowers being
so strongly proterandrous. From
what Lecoq says, *S. succisa* likewise appears to occur under two
forms in France.

are short; the anthers do not contain any sound pollen-grains, but in their place yellow incoherent cells which do not swell in water. Some plants were in an intermediate condition; that is, had one or two or three stamens of proper length with perfect anthers, the other stamens being rudimentary. In one such plant half of one anther contained green perfect pollen-grains, and the other half yellowish-green imperfect grains. Both forms produced seed, but I neglected to observe whether in equal numbers. As I thought that the state of the anthers might be due to some fungoid growth, I examined them both in the bud and mature state, but could find no trace of mycelium. In 1862 many female plants were found; and in 1864, 32 plants were collected in two localities, exactly half of which were hermaphrodites, fourteen were females, and two in an intermediate condition. In 1866, 15 plants were collected in another locality, and these consisted of four hermaphrodites and eleven females. I may add that this season was a wet one, which shows that the abortion of the stamens can hardly be due to the dryness of the sites where the plants grew, as I at one time thought probable. Seeds from an hermaphrodite were sown in my garden, and of the 23 seedlings raised, one belonged to the intermediate form, all the others being hermaphrodites, though two or three of them had unusually short stamens. I have consulted several botanical works, but have found no record of this plant varying in the manner here described.

Plantago lanceolata (Plantagineæ)—Delpino states that this plant presents in Italy three forms, which graduate from an anemophilous into an entomophilous condition. According to H. Müller,[*] there are only two forms in Germany, neither of which show any special adaptation for insect fertilisation, and both appear to be hermaphrodites. But I have found in two localities in England female and hermaphrodite forms existing together; and the same fact has been noticed by others.[†] The females are less frequent than the hermaphrodites; their stamens are short, and their anthers, which are of a brighter green whilst young than those of the other form, dehisce properly, yet contain either no pollen, or a small amount of imperfect grains of variable size. All the flower-heads on a plant belong to

---

[*] 'Die Befruchtung,' &c., p. 342.
[†] Mr. C. W. Crocker in 'The Gardener's Chronicle,' 1864, p. 294.

Mr. W. Marshall writes to me to the same effect from Ely.

the same form. It is well known that this species is strongly proterogynous, and I found that the protruding stigmas of both the hermaphrodite and female flowers were penetrated by pollen-tubes, whilst their own anthers were immature and had not escaped out of the bud. *Plantago media* does not present two forms; but it appears from Asa Gray's description,* that such is the case with four of the North American species. The corolla does not properly expand in the short-stamened form of these plants.

Cnicus, Serratula, Eriophorum.—In the Compositæ, *Cnicus palustris* and *acaulis* are said by Sir J. E. Smith to exist as hermaphrodites and females, the former being the more frequent. With *Serratula tinctoria* a regular gradation may be followed from the hermaphrodite to the female form; in one of the latter plants the stamens were so tall that the anthers embraced the style as in the hermaphrodites, but they contained only a few grains of pollen, and these in an aborted condition; in another female, on the other hand, the anthers were much more reduced in size than is usual. Lastly, Dr. Dickie has shown that with *Eriophorum angustifolium* (Cyperaceæ) hermaphrodite and female forms exist in Scotland and the Arctic regions, both of which yield seed.†

It is a curious fact that in all the foregoing polygamous, diœcious, and gyno-diœcious plants in which any difference has been observed in the size of the corolla in the two or three forms, it is rather larger in the females, which have their stamens more or less or quite rudimentary, than in the hermaphrodites or males. This holds good with Euonymus, *Rhamnus catharticus*, Ilex, Fragaria, all or at least most of the before-named Labiatæ, *Scabiosa atro-purpurea*, and *Echium vulgare*. So it is, according to Von Mohl, with *Cardamine*

---

* 'Manual of the Botany of the N. United States,' 2nd edit. 1856, p. 269. See also 'American Journal of Science,' Nov. 1862, p. 419, and 'Proc. American Academy of Science,' Oct. 14, 1862, p. 53.

† Sir J. E. Smith, ' Trans. Linn. Soc.' vol. xiii. p. 599. Dr. Dickie, 'Journal Linn. Soc. Bot.' vol. ix. 1865, p. 161.

amara, Geranium sylvaticum, Myosotis, and *Salvia.*
On the other hand, as Von Mohl remarks, when a
plant produces hermaphrodite flowers and others
which are males owing to the more or less complete
abortion of the female organs, the corollas of the
males are not at all increased in size, or only excep-
tionally and in a slight degree, as in Acer. * It seems
therefore probable that the decreased size of the female
corollas in the foregoing cases is due to a tendency to
abortion spreading from the stamens to the petals. We
see how intimately these organs are related in double
flowers, in which the stamens are readily converted
into petals. Indeed some botanists believe that petals
do not consist of leaves directly metamorphosed, but of
metamorphosed stamens. That the lessened size of the
corolla in the above case is in some manner an indirect
result of the modification of the reproductive organs is
supported by the fact that in *Rhamnus catharticus* not
only the petals but the green and inconspicuous sepals
of the female have been reduced in size; and in the
strawberry the flowers are largest in the males, mid-
sized in the hermaphrodites, and smallest in the females. These latter cases,—the variability in the size
of the corolla in some of the above species, for instance
in the common thyme,—together with the fact that it
never differs greatly in size in the two forms—make
me doubt much whether natural selection has come
into play;—that is whether, in accordance with H.
Müller's belief, the advantage derived from the polleniferous flowers being visited first by insects has
been sufficient to lead to a gradual reduction of the
corolla of the female. We should bear in mind that as
the hermaphrodite is the normal form, its corolla has

---

* 'Bot. Zeitung,' 1863, p. 326.

probably retained its original size.*   An objection to
the above view should not be passed over; namely, that
the abortion of the stamens in the females ought to
have added through the law of compensation to the
size of the corolla; and this perhaps would have occurred, had not the expenditure saved by the abortion of the stamens been directed to the female reproductive organs, so as to give to this form increased
fertility.

---

* It does not appear to me
that Kerner's view ('Die Schutzmittel des Pollens,' 1873, p. 56)
can be accepted in the present
cases, namely that the larger
corolla in the hermaphrodites and
males serves to protect their pollen
from rain.   In the genus Thymus,
for instance, the aborted anthers
of the female are much better
protected than the perfect ones of
the hermaphrodite.

CHAPTER VIII.

Cleistogamic Flowers.

General character of cleistogamic flowers—List of the genera producing such flowers, and their distribution in the vegetable series—Viola, description of the cleistogamic flowers in the several species, their fertility compared with that of the perfect flowers—Oxalis acetosella—O. sensitiva, three forms of cleistogamic flowers—Vandellia—Ononis—Impatiens—Drosera—Miscellaneous observations on various other cleistogamic plants—Anemophilous species producing cleistogamic flowers—Leersia, perfect flowers rarely developed—Summary and concluding remarks on the origin of cleistogamic flowers—The chief conclusions which may be drawn from the observations in this volume.

It was known even before the time of Linnæus that certain plants produced two kinds of flowers, ordinary open, and minute closed ones; and this fact formerly gave rise to warm controversies about the sexuality of plants. These closed flowers have been appropriately named cleistogamic by Dr. Kuhn.[*] They are remarkable from their small size and from never opening, so that they resemble buds; their petals are rudimentary or quite aborted; their stamens are often reduced in number, with the anthers of very small size, containing few pollen-grains, which have remarkably thin transparent coats, and generally emit their tubes whilst still enclosed within the anther-cells; and, lastly, the pistil is much reduced in size, with the stigma in some cases hardly at all developed. These flowers do not secrete nectar or emit any odour; from their small size, as well as from the corolla being rudimentary, they are singularly inconspicuous.   Con-

---

[*] ' Bot. Zeitung,' 1867, p. 65.

sequently insects do not visit them; nor if they did, could they find an entrance. Such flowers are therefore invariably self-fertilised; yet they produce an abundance of seed. In several cases the young capsules bury themselves beneath the ground, and the seeds are there matured. These flowers are developed before, or after, or simultaneously with the perfect ones. Their development seems to be largely governed by the conditions to which the plants are exposed, for during certain seasons or in certain localities only cleistogamic or only perfect flowers are produced.

Dr. Kuhn, in the article above referred to, gives a list of 44 genera including species which bear flowers of this kind. To this list I have added some genera, and the authorities are appended in a foot-note. I have omitted three names, from reasons likewise given in the foot-note. But it is by no means easy to decide in all cases whether certain flowers ought to be ranked as cleistogamic. For instance, Mr. Bentham informs me that in the South of France some of the flowers on the vine do not fully open and yet set fruit; and I hear from two experienced gardeners that this is the case with the vine in our hot-houses; but as the flowers do not appear to be completely closed it would be imprudent to consider them as cleistogamic. The flowers of some aquatic and marsh plants, for instance of *Ranunculus aquatilis, Alisma natans,* Subularia, Illecebrum, Menyanthes, and Euryale,[*] remain closely shut as long as they are submerged, and in this condition fertilise themselves.

---

[*] Delpino, 'Sull' Opera, la Distribuzione dei Sessi nelle Piante,' &c. 1867, p. 30. Subularia, however, sometimes has its flowers fully expanded beneath the water, see Sir J. E. Smith, 'English Flora,' vol. iii. 1825, p. 157. For the behaviour of Menyanthes in Russia see Gillibert in 'Act. Acad. St. Petersb.,' 1777, part ii. p. 45.—On Euryale, 'Gardener's Chronicle,' 1877, p. 280.

They behave in this manner, apparently as a protection to their pollen, and produce open flowers when exposed to the air; so that these cases seem rather different from those of true cleistogamic flowers, and have not been included in the list.   Again, the flowers of some plants which are produced very early or very late in the season do not properly expand; and these might perhaps be considered as incipiently cleistogamic; but as they do not present any of the remarkable peculiarities proper to the class, and as I have not found any full record of such cases, they are not entered in the list.   When, however, it is believed on fairly good evidence that the flowers on a plant in its native country do not open at any hour of the day or night, and yet set seeds capable of germination, these may fairly be considered as cleistogamic, notwithstanding that they present no peculiarities of structure.   I will now give as complete a list of the genera containing cleistogamic species as I have been able to collect.

TABLE 38.

List of Genera including Cleistogamic Species (chiefly after Kuhn).*

DICOTYLEDONS.	DICOTYLEDONS.
Eritrichium (Boragineæ).	Dædalacanthus (Acanthaceæ).
Cuscuta (Convolvulaceæ).	Dipteracanthus „
Scrophularia (Scrophularineæ).	Æchmanthera „
Linaria „	Ruellia „
Vandellia „	Lamium (Labiatæ).
Cryphiacanthus (Acanthaceæ).	Salvia „
Eranthemum „	Oxybaphus (Nyctagineæ).

* I have omitted Trifolium and Arachis from the list, because Von Mohl says ('Bot. Zeitung,' 1863, p. 312) that the flower-stems merely draw the flowers beneath the ground, and that these do not appear to be properly cleistogamic. Correa de Mello ('Journal Linn. Soc. Bot.' vol. xi. 1870, p. 254) observed plants of Arachis in Brazil, and could never find such flowers. Plantago has been omitted because as far as I can discover it produces hermaphrodite and female flower-heads, but not cleistogamic flowers. Krascheninikowia (vel Stellaria) has been omitted because it seems very doubtful from Maximowicz' description whether the lower flowers which

Table 38—*continued.*

DICOTYLEDONS.

Nyctaginia (Nyctagineæ).
Stapelia (Asclepiadæ).
Specularia (Campanulaceæ).
Campanula „
Hottonia (Primulaceæ).
Anandria (Compositæ).
Heterocarpæa (Cruciferæ).
Viola (Violaceæ).
Helianthemum (Cistineæ).
Lechea „
Pavonia (Malvaceæ).
Gaudichaudia (Malpighiaceæ).
Aspicarpa „
Camarea „
Janusia „
Polygala (Polygaleæ).
Impatiens (Balsamineæ).
Oxalis (Geraniaceæ).
Ononis (Leguminosæ).
Parochætus „
Chapmannia „
Stylosanthus „

DICOTYLEDONS.

Lespedeza (Leguminosæ).
Vicia „
Lathyrus „
Martiusia *vel* }
Neurocarpum } „
Amphicarpæa „
Glycine „
Galactia „
Voandzeia „
Drosera (Droseraceæ).

MONOCOTYLEDONS.

Juncus (Junceæ).
Leersia (Gramineæ).
Hordeum „
Cryptostachys „
Commelina (Commelineæ).
Monochoria (Pontederaceæ).
Schomburgkia (Orchideæ).
Cattleya „
Epidendron „
Thelymitra „

have no petals or very small ones, and barren stamens or none, are cleistogamic ; the upper hermaphrodite flowers are said never to produce fruit, and therefore probably act as males. Moreover in *Stellaria graminea,* as Babington remarks ('British Botany,' 1851, p. 51) "shorter and longer petals accompany an imperfection of the stamens or germen."

I have added to the list the following cases : Several Acanthaceæ, for which see J. Scott in 'Journal of Bot.' (London), new series, vol. i. 1872, p. 161. With respect to Salvia see Dr. Ascherson in 'Bot. Zeitung,' 1871, p. 555. For Oxybaphus and Nyctaginia see Asa Gray in 'American Naturalist,' Nov. 1873, p. 692. From Dr. Torrey's account of *Hottonia inflata* ('Bull. of Torrey Botan. Club,' vol. ii. June 1871) it is manifest that this plant produces true cleistogamic flowers. For

Pavonia see Bouché in 'Sitzungsberichte d. Gesellsch. Natur. Freunde,' Oct. 20, 1874, p. 90. I have added Thelymitra,as from the account given by Mr. Fitzgerald in his magnificent work on 'Australian Orchids' it appears that the flowers of this plant in its native home never open, but they do not appear to be reduced in size. Nor is this the case with the flowers of certain species of Epidendron, Cattleya, &c. (see second edition of my 'Fertilisation of Orchids,' p. 147), which without expanding produce capsules. It is therefore doubtful whether these Orchideæ ought to have been included in the list. From what Duval-Jouve says about Cryptostachys in 'Bull. Soc. Bot. de France,' tom. x. 1863, p. 195, this plant appears to produce cleistogamic flowers. The other additions to the list are noticed in my text.

The first point that strikes us in considering this list of 55 genera, is that they are very widely distributed in the vegetable series. They are more common in the family of the Leguminosæ than in any other, and next in order in that of the Acanthaceæ and Malpighiaceæ. A large number, but not all the species, of certain genera, as of Oxalis and Viola, bear cleistogamic as well as ordinary flowers. A second point which deserves notice is that a considerable proportion of the genera produce more or less irregular flowers; this is the case with about 32 out of the 55 genera, but to this subject I shall recur.

I formerly made many observations on cleistogamic flowers, but only a few of them are worth giving, since the appearance of an admirable paper by Hugo von Mohl,* whose examination was in some respects much more complete than mine. His paper includes also an interesting history of our knowledge on the subject.

Viola canina.—The calyx of the cleistogamic flowers differs in no respect from that of the perfect ones. The petals are reduced to five minute scales; the lower one, which represents the lower lip, is considerably larger than the others, but with no trace of the spur-like nectary; its margins are smooth, whilst those of the other four scale-like petals are papillose. D. Müller of Upsala says that in the specimens which he observed the petals were completely aborted.† The stamens are very small, and only the two lower ones are provided with anthers, which do not cohere together as in the perfect flowers. The anthers are minute, with the two cells or loculi remarkably distinct; they contain very little pollen in comparison with those of the perfect

---

* 'Bot. Zeitung,' 1863, p. 309–28.

† Ibid. 1857, p. 730. This paper contains the first full and satisfactory account of any cleistogamic flower.

flowers. The connective expands into a membranous hood-like shield which projects above the anther-cells. These two lower stamens have no vestige of the curious appendages which secrete nectar in the perfect flowers. The three other stamens are destitute of anthers and have broader filaments, with their terminal membranous expansions flatter or not so hood-like as those of the two antheriferous stamens. The pollen-grains have remarkably thin transparent coats; when exposed to the air they shrivel up quickly; when placed in water they swell, and are then $\frac{8-10}{7000}$ of an inch in diameter, and therefore of smaller size than the ordinary pollen-grains similarly treated, which have a diameter of $\frac{13-14}{7000}$ of an inch. In the cleistogamic flowers, the pollen-grains, as far as I could see, never naturally fall out of the anther-cells, but emit their tubes through a pore at the upper end. I was able to trace the tubes from the grains some way down the stigma. The pistil is very short, with the style hooked, so that its extremity, which is a little enlarged or funnel-shaped and represents the stigma, is directed downwards, being covered by the two membranous expansions of the antheriferous stamens. It is remarkable that there is an open passage from the enlarged funnel-shaped extremity to within the ovarium; this was evident, as slight pressure caused a bubble of air, which had been drawn in by some accident, to travel freely from one end to the other : a similar passage was observed by Michalet in *V. alba*. The pistil therefore differs considerably from that of the perfect flower; for in the latter it is much longer, and straight with the exception of the rectangularly bent stigma; nor is it perforated by an open passage.

The ordinary or perfect flowers have been said by some authors never to produce capsules; but this is an error, though only a small proportion of them do so. This appears to depend in some cases on their anthers not containing even a trace of pollen, but more generally on bees not visiting the flowers. I twice covered with a net a group of flowers, and marked with threads twelve of them which had not as yet expanded. This precaution is necessary, for though as a general rule the perfect flowers appear considerably before the cleistogamic ones, yet occasionally some of the latter are produced early in the season, and their capsules might readily be mistaken for those produced by the perfect flowers. Not one of the twelve marked perfect flowers yielded a capsule, whilst others under the net which had been artificially fertilised produced five capsules ; and these contained exactly the same average number of seeds as some capsules from flowers outside the net which had been fertilised by bees. I have repeatedly seen *Bombus hortorum, lapidarius,* and a third species, as well as hive-bees, sucking the flowers of this violet : I marked six which were thus visited, and four of them produced fine capsules ; the two others were gnawed off by some animal. I watched *Bombus hortorum* for some time, and whenever it came to a flower which did not stand in a convenient position to be sucked, it bit a hole through the spur-like nectary. Such ill-placed flowers would not yield any seed or leave descendants ; and the plants bearing them would thus tend to be eliminated through natural selection.

The seeds produced by the cleistogamic and perfect flowers do not differ in appearance or number. On two occasions I fertilised several perfect flowers with pollen from other individuals, and afterwards marked some cleistogamic flowers on the same plants ; and the result was that 14 capsules produced by the perfect flowers contained on an average 9·85 seeds; and 17 capsules from the cleistogamic ones contained 9·64 seeds,—an amount of difference of no significance. It is remarkable how much more quickly the capsules from the cleistogamic flowers are developed than those from the perfect ones; for instance, several perfect flowers were cross-fertilised on April 14th, 1863, and a month afterwards (May 15th) eight young cleistogamic flowers were marked with threads; and when the two sets of capsules thus produced were compared on June 3rd, there was scarcely any difference between them in size.

Viola odorata (white-flowered, single, cultivated variety).—The petals are represented by mere scales as in the last species; but differently from in the last, all five stamens are provided with diminutive anthers. Small bundles of pollen-tubes were traced from the five anthers into the somewhat distant stigma. The capsules produced by these flowers bury themselves in the soil, if it be loose enough, and there mature themselves.* Lecoq says that it is only these latter capsules which possess elastic valves; but I think this must be a misprint, as such valves would obviously be of no use to the buried capsules, but would serve to scatter the seeds of the sub-aerial ones, as in the other species of Viola. It is remarkable that this plant, according to Delpino,† does not produce cleistogamic flowers in one part of Liguria, whilst the perfect flowers are there abundantly fertile; on the other hand, cleistogamic flowers are produced by it near Turin. Another fact is worth giving as an instance of corre-

---

* Vaucher says ('Hist. Phys. des Plantes d'Europe,' tom. iii. 1844, p. 309) that *V. hirta* and *collina* likewise bury their capsules.— See also Lecoq, 'Géograph. Bot.'

tom. v. 1856, p. 180.
   † 'Sull' Opera, la Distribuzione dei Sessi nelle Piante,' &c., 1867, p. 30.

lated development : I found on a purple variety, after it had produced its perfect double flowers, and whilst the white single variety was bearing its cleistogamic flowers, many bud-like bodies which from their position on the plant were certainly of a cleistogamic nature. They consisted, as could be seen on bisecting them, of a dense mass of minute scales closely folded over one another, exactly like a cabbage-head in miniature. I could not detect any stamens, and in the place of the ovarium there was a little central column. The doubleness of the perfect flowers had thus spread to the cleistogamic ones, which therefore were rendered quite sterile.

Viola hirta.—The five stamens of the cleistogamic flowers are provided, as in the last case, with small anthers, from all of which pollen-tubes proceed to the stigma. The petals are not quite so much reduced as in *V. canina,* and the short pistil instead of being hooked is merely bent into a rectangle. Of several perfect flowers which I saw visited by hive- and humble-bees, six were marked, but they produced only two capsules, some of the others having been accidentally injured. M. Monnier was therefore mistaken in this case as in that of *V. odorata,* in supposing that the perfect flowers always withered away and aborted. He states that the peduncles of the cleistogamic flowers curve downwards and bury the ovaries beneath the soil.* I may here add that Fritz Müller, as I hear from his brother, has found in the highlands of Southern Brazil a white-flowered species of violet which bears subterranean cleistogamic flowers.

---

* These statements are taken from Professor Oliver's excellent article in the ' Nat. Hist. Review,' July 1862, p. 238. With respect to the supposed sterility of the perfect flowers in this genus see also Timbal-Lagrave in ' Bot. Zeitung,' 1854, p. 772.

Viola nana.—Mr. Scott sent me seeds of this Indian species from the Sikkim Terai, from which I raised many plants, and from these other seedlings during several successive generations. They produced an abundance of cleistogamic flowers during the whole of each summer, but never a perfect one. When Mr. Scott wrote to me his plants in Calcutta were behaving similarly, though his collector saw the species in flower in its native site. This case is valuable as showing that we ought not to infer, as has sometimes been done, that a species does not bear perfect flowers when growing naturally, because it produces only cleistogamic flowers under culture. The calyx of these flowers is sometimes formed of only three sepals; two being actually suppressed and not merely coherent with the others; this occurred with five out of thirty flowers which were examined for this purpose. The petals are represented by extremely minute scales. Of the stamens, two bear anthers which are in the same state as in the previous species, but, as far as I could judge, each of the two cells contained only from 20 to 25 delicate transparent pollen-grains. These emitted their tubes in the usual manner. The three other stamens bore very minute rudimentary anthers, one of which was generally larger than the other two, but none of them contained any pollen. In one instance, however, a single cell of the larger rudimentary anther included a little pollen. The style consists of a short flattened tube, somewhat expanded at its upper end, and this forms an open channel leading into the ovarium, as described under *V. canina.* It is slightly bent towards the two fertile anthers.

Viola Roxburghiana.—This species bore in my hothouse during two years a multitude of cleistogamic flowers, which resembled in all respects those of the last species; but no perfect ones were produced. Mr. Scott informs me that in India it bears perfect flowers only during the cold season, and that these are quite fertile. During the hot, and more especially during the rainy season, it bears an abundance of cleistogamic flowers.

Many other species, besides the five now described, produce cleistogamic flowers; this is the case, according to D. Müller, Michalet, Von Mohl, and Hermann Müller, with *V. elatior, lancifolia, sylvatica, palustris, mirabilis, bicolor, ionodium,* and *biflora.* But *V. tricolor* does not produce them.

Michalet asserts that *V. palustris* produces near Paris only perfect flowers, which are quite fertile; but that when the plant grows on mountains cleistogamic flowers are produced; and so it is with *V. biflora.* The same author states that he has seen in the case of *V. alba* flowers intermediate in structure between the perfect and cleistogamic ones. According to M. Boisduval, an Italian species, *V. Ruppii,* never bears in France " des fleurs bien apparentes, ce qui ne l'empêche pas de fructifier."

It is interesting to observe the gradation in the abortion of the parts in the cleistogamic flowers of the several foregoing species. It appears from the statements by D. Müller and Von Mohl that in *V. mirabilis* the calyx does not remain quite closed; all five stamens are provided with anthers, and some pollengrains probably fall out of the cells on the stigma, instead of protruding their tubes whilst still enclosed, as in the other species. In *V. hirta* all five stamens are likewise antheriferous; the petals are not so much reduced and the pistil not so much modified as in the following species. In *V. nana* and *elatior* only two of the stamens properly bear anthers, but

sometimes one or even two of the others are thus provided. Lastly, in *V. canina* never more than two of
the stamens, as far as I have seen, bear anthers; the
petals are much more reduced than in *V. hirta*, and
according to D. Müller are sometimes quite absent.

Oxalis acetosella.—The existence of cleistogamic
flowers on this plant was discovered by Michalet.*
They have been fully described by Von Mohl, and I
can add hardly anything to his description. In my
specimens the anthers of the five longer stamens were
nearly on a level with the stigmas; whilst the smaller
and less plainly bilobed anthers of the five shorter
stamens stood considerably below the stigmas, so that
their tubes had to travel some way upwards. According to Michalet these latter anthers are sometimes quite aborted. In one case the tubes, which
ended in excessively fine points, were seen by me
stretching upwards from the lower anthers towards
the stigmas, which they had not as yet reached. My
plants grew in pots, and long after the perfect flowers
had withered they produced not only cleistogamic but
a few minute open flowers, which were in an intermediate condition between the two kinds. In one of
these the pollen-tubes from the lower anthers had
reached the stigmas, though the flower was open.
The footstalks of the cleistogamic flowers are much
shorter than those of the perfect flowers, and are so
much bowed downwards that they tend, according to
Von Mohl, to bury themselves in the moss and dead
leaves on the ground. Michalet also says that they
are often hypogean. In order to ascertain the number of seeds produced by these flowers, I marked eight
of them; two failed, one cast its seed abroad, and the

---

* 'Bull. Soc. Bot. de France,' tom. vii. 1860, p. 465.

remaining five contained on an average 10·0 seeds
per capsule. This is rather above the average 9·2,
which eleven capsules from perfect flowers fertilised
with their own pollen yielded, and considerably above
the average 7·9, from the capsules of perfect flowers
fertilised with pollen from another plant; but this
latter result must, I think, have been accidental.

Hildebrand, whilst searching various Herbaria, observed that many other species of Oxalis besides *O.
acetosella* produce cleistogamic flowers;* and I hear
from him that this is the case with the heterostyled
trimorphic *O. incarnata* from the Cape of Good Hope.

Oxalis (Biophytum) sensitiva.—This plant is ranked
by many botanists as a distinct genus, but as a sub-
genus by Bentham and Hooker. Many of the early
flowers on a mid-styled plant in my hot-house did not
open properly, and were in an intermediate condition
between cleistogamic and perfect. Their petals varied
from a mere rudiment to about half their proper size;
nevertheless they produced capsules. I attributed
their state to unfavourable conditions, for later in the
season fully expanded flowers of the proper size appeared. But Mr. Thwaites afterwards sent me from
Ceylon a number of long-styled, mid-styled, and short-
styled flower-stalks preserved in spirits; and on the
same stalks with the perfect flowers, some of which were
fully expanded and others still in bud, there were
small bud-like bodies containing mature pollen, but
with their calyces closed. These cleistogamic flowers
do not differ much in structure from the perfect ones
of the corresponding form, with the exception that
their petals are reduced to extremely minute, barely
visible scales, which adhere firmly to the rounded

---

* 'Monatsbericht der Akad. der Wiss. zu Berlin,' 1866, p. 369.

bases of the shorter stamens. Their stigmas are much
less papillose, and smaller in about the ratio of 13 to
20 divisions of the micrometer, as measured trans-
versely from apex to apex, than the stigmas of the
perfect flowers. The styles are furrowed longitudinally,
and are clothed with simple as well as glandular hairs,
but only in the cleistogamic flowers produced by the
long-styled and mid-styled forms. The anthers of the
longer stamens are a little smaller than the corresponding ones of the perfect flowers, in about the ratio of
11 to 14. They dehisce properly, but do not appear
to contain much pollen. Many pollen-grains were
attached by short tubes to the stigmas ; but many
others, still adhering to the anthers, had emitted
their tubes to a considerable length, without having
come in contact with the stigmas. Living plants
ought to be examined, as the stigmas, at least of the
long-styled form, project beyond the calyx, and if
visited by insects (which, however, is very improbable)
might be fertilised with pollen from a perfect flower.
The most singular fact about the present species is
that long-styled cleistogamic flowers are produced by
the long-styled plants, and mid-styled as well as
short-styled cleistogamic flowers by the other two
forms ; so that there are three kinds of cleistogamic
and three kinds of perfect flowers produced by this
one species ! Most of the heterostyled species of
Oxalis are more or less sterile, many absolutely so, if
illegitimately fertilised with their own-form pollen.
It is therefore probable that the pollen of the cleistogamic flowers has been modified in power, so as to act
on their own stigmas, for they yield an abundance of
seeds. We may perhaps account for the cleistogamic
flowers consisting of the three forms, through the principle of correlated growth, by which the cleistogamic

flowers of the double violet have been rendered double.

Vandellia nummularifolia.—Dr. Kuhn has collected[*] all the notices with respect to cleistogamic flowers in this genus, and has described from dried specimens those produced by an Abyssinian species. Mr. Scott sent me from Calcutta seeds of the above common Indian weed, from which many plants were successively raised during several years. The cleistogamic flowers are very small, being when fully mature under $\frac{1}{20}$ of an inch (1·27 mm.) in length. The calyx does not open, and within it the delicate transparent corolla remains closely folded over the ovarium. There are only two anthers instead of the normal number of four, and their filaments adhere to the corolla. The cells of the anthers diverge much at their lower ends and are only $\frac{5}{700}$ of an inch (·181 mm.) in their longer diameter. They contain but few pollen-grains, and these emit their tubes whilst still within the anther. The pistil is very short, and is surmounted by a bilobed stigma. As the ovary grows the two anthers together with the shrivelled corolla, all attached by the dried pollen-tubes to the stigma, are torn off and carried upwards in the shape of a little cap. The perfect flowers generally appear before the cleistogamic, but sometimes simultaneously with them. During one season a large number of plants produced no perfect flowers. It has been asserted that the latter never yield capsules; but this is a mistake, as they do so even when insects are excluded. Fifteen capsules from cleistogamic flowers on plants growing under favourable conditions contained on an average 64·2 seeds, with a maximum of 87; whilst 20 capsules from plants growing much

---

[*] 'Bot. Zeitung,' 1867, p. 65.

crowded yielded an average of only 48. Sixteen capsules from perfect flowers artificially crossed with pollen
from another plant contained on an average 93 seeds,
with a maximum of 137. Thirteen capsules from self-
fertilised perfect flowers gave an average of 62 seeds,
with a maximum of 135. Therefore the capsules from
the cleistogamic flowers contained fewer seeds than
those from perfect flowers when cross-fertilised, and
slightly more than those from perfect flowers self-
fertilised.

Dr. Kuhn believes that the Abyssinian *V. sessiflora*
does not differ specifically from the foregoing species.
But its cleistogamic flowers apparently include four
anthers instead of two as above described. The plants,
moreover, of *V. sessiflora* produce subterranean runners
which yield capsules; and I never saw a trace of such
runners in *V. nummularifolia*, although many plants
were cultivated.

Linaria spuria.—Michalet says* that short, thin,
twisted branches are developed from the buds in the
axils of the lower leaves, and that these bury themselves in the ground. They there produce flowers
not offering any peculiarity in structure, excepting
that their corollas, though properly coloured, are de-
formed. These flowers may be ranked as cleistogamic,
as they are developed, and not merely drawn, beneath
the ground.

Ononis columnæ.—Plants were raised from seeds sent
me from Northern Italy. The sepals of the cleistogamic flowers are elongated and closely pressed together; the petals are much reduced in size, colourless, and folded over the interior organs. The filaments of the ten stamens are united into a tube, and

---

* 'Bull. Soc. Bot. de France,' tom. vii. 1860, p. 468.

this is not the case, according to Von Mohl, with the
cleistogamic flowers of other Leguminosæ.   Five of
the stamens are destitute of anthers, and alternate with
the five thus provided.   The two cells of the anthers
are minute, rounded and separated from one another
by connective tissue; they contain but few pollen-
grains, and these have extremely delicate coats.   The
pistil is hook-shaped, with a plainly enlarged stigma,
which is curled down, towards the anthers; it therefore differs much from that of the perfect flower.
During the year 1867 no perfect flowers were produced, but in the following year there were both
perfect and cleistogamic ones.

Ononis minutissima.—My plants produced both perfect and cleistogamic flowers; but I did not examine
the latter.   Some of the former were crossed with
pollen from a distinct plant, and six capsules thus obtained yielded on an average 3·66 seeds, with a maximum of 5 in one.   Twelve perfect flowers were marked
and allowed to fertilise themselves spontaneously under
a net, and they yielded eight capsules, containing on
an average 2·38 seeds, with a maximum of 3 in one.
Fifty-three capsules produced by the cleistogamic
flowers contained on an average 4·1 seeds, so that
these were the most productive of all; and the seeds
themselves looked finer even than those from the
crossed perfect flowers.   According to Mr. Bentham
O. parviflora likewise bears cleistogamic flowers; and
he informs me that these flowers are produced by all
three species early in the spring; whilst the perfect
ones appear afterwards, and therefore in a reversed
order compared with those of Viola and Oxalis.   Some
of the species, for instance *Ononis columnæ*, bear a
fresh crop of cleistogamic flowers in the autumn.

Lathyrus nissolia apparently offers a case of the first

stage in the production of cleistogamic flowers, for on plants growing in a state of nature, many of the flowers never expand and yet produce fine pods. Some of the buds are so large that they seem on the point of expansion ; others are much smaller, but none so small as the true cleistogamic flowers of the foregoing species. As I marked these buds with thread and examined them daily, there could be no mistake about their producing fruit without having expanded.

Several other Leguminous genera produce cleistogamic flowers, as may be seen in the previous list; but much does not appear to be known about them. Von Mohl says that their petals are commonly rudimentary, that only a few of their anthers are developed, their filaments are not united into a tube and their pistils are hook-shaped. In three of the genera, namely Vicia, Amphicarpæa, and Voandzeia, the cleistogamic flowers are produced on subterranean stems. The perfect flowers of Voandzeia, which is a cultivated plant, are said never to produce fruit ;* but we should remember how often fertility is affected by cultivation.

Impatiens fulva.—Mr. A. W. Bennett has published an excellent description, with figures, of this plant.† He shows that the cleistogamic and perfect flowers differ in structure at a very early period of growth, so that the existence of the former cannot be due merely to the arrested development of the latter,—a conclusion which indeed follows from most of the previous descriptions. Mr. Bennett found on the banks of the Wey that the plants which bore cleistogamic flowers alone were to those bearing perfect flowers as 20 to 1 ; but

---

* Correa de Mello ('Journal Linn. Soc. Bot.' vol. xi. 1870, p. 254) particularly attended to the flowering and fruiting of this African plant, which is sometimes cultivated in Brazil.

† 'Journal Linn. Soc. Bot.' vol. xiii. 1872, p. 147.

we should remember that this is a naturalised species.
The perfect flowers are usually barren in England; but
Prof. Asa Gray writes to me that after midsummer in
the United States some or many of them produce
capsules.

Impatiens noli-me-tangere.—I can add nothing of importance to Von Mohl's description, excepting that
one of the rudimentary petals shows a vestige of a
nectary, as Mr. Bennett likewise found to be the case
with *I. fulva*. As in this latter species all five stamens
produce some pollen, though small in amount; a
single anther contains, according to Von Mohl, not
more than 50 grains, and these emit their tubes
while still enclosed within it. The pollen-grains of the
perfect flowers are tied together by threads, but not,
so as far as I could see, those of the cleistogamic
flowers; and a provision of this kind would here have
been useless, as the grains can never be transported
by insects. The flowers of *I. balsamina* are visited by
humble-bees,* and I am almost sure that this is the
case with the perfect flowers of *I. noli-me-tangere*. From
the perfect flowers of this latter species covered with
a net eleven spontaneously self-fertilised capsules were
produced, and these yielded on an average 3·45 seeds.
Some perfect flowers with their anthers still containing
an abundance of pollen were fertilised with pollen from
a distinct plant; and the three capsules thus produced
contained, to my surprise, only 2, 2, and 1 seed. As
I. balsamina is proterandrous, so probably is the present species; and if so, cross-fertilisation was effected
by me at too early a period, and this may account for
the capsules yielding so few seeds.

Drosera rotundifolia.—The first flower-stems which

---

* H. Müller, 'Die Befruchtung,' &c. p. 170.

were thrown up by some plants in my green-house bore only cleistogamic flowers. The petals of small size remained permanently closed over the reproductive organs, but their white tips could just be seen between the almost completely closed sepals. The pollen, which was scanty in amount, but not so scanty as in Viola or Oxalis, remained enclosed within the anthers, whence the tubes proceeded and penetrated the stigma. As the ovarium swelled the little withered corolla was carried upwards in the form of a cap. These cleistogamic flowers produced an abundance of seed. Later in the season perfect flowers appeared. With plants in a state of nature the flowers open only in the early morning, as I have been informed by Mr. Wallis, who particularly attended to the time of their flowering. In the case of *D. Anglica*, the still folded petals on some plants in my green-house opened just sufficiently to leave a minute aperture; the anthers dehisced properly, but the pollen-grains adhered in a mass to them, and thence emitted their tubes, which penetrated the stigmas. These flowers, therefore, were in an intermediate condition, and could not be called either perfect or cleistogamic.

A few miscellaneous observations may be added with respect to some other species, as throwing light on our subject. Mr. Scott states * that *Eranthemum ambiguum* bears three kinds of flowers,—large, conspicuous, open ones, which are quite sterile,—others of intermediate size, which are open and moderately fertile—and lastly small closed or cleistogamic ones, which are perfectly fertile. *Ruellia tuberosa*, likewise one of the Acanthaceæ, produces both open and cleis-

---

* 'Journal of Botany,' London, new series, vol. i. 1872, pp. 161-4.

togamic flowers; the latter yield from 18 to 24, whilst
the former only from 8 to 10 seeds; these two kinds of
flowers are produced simultaneously, whereas in several
other members of the family the cleistogamic ones
appear only during the hot season. According to
Torrey and Gray, the North American species of Helianthemum, when growing in poor soil, produce only
cleistogamic flowers. The cleistogamic flowers of
Specularia perfoliata are highly remarkable, as they
are closed by a tympanum formed by the rudimentary corolla, and without any trace of an opening. The stamens vary from 3 to 5 in number,
as do the sepals.* The collecting hairs on the pistil,
which play so important a part in the fertilisation
of the perfect flowers, are here quite absent. Drs.
Hooker and Thomson state† that some of the Indian
species of Campanula produce two kinds of flowers;
the smaller ones being borne on longer peduncles
with differently formed sepals, and producing a more
globose ovary. The flowers are closed by a tympanum like that in Specularia. Some of the plants
produce both kinds of flowers, others only one kind;
both yield an abundance of seeds. Professor Oliver
adds that he has seen flowers on *Campanula colorata*
in an intermediate condition between cleistogamic and
perfect ones.

The solitary almost sessile cleistogamic flowers produced by *Monochoria vaginalis* are differently protected
from those in any of the previous cases, namely, within
" a short sack formed of the membranous spathe,

---

* Von Mohl, 'Bot. Zeitung,'
1863, pp. 314 and 323. Dr. Bromfield ('Phytologist,' vol. iii. p.
530) also remarks that the calyx
of the cleistogamic flowers is
usually only 3-cleft, while that

of the perfect flower is mostly
5-cleft.
    † 'Journal Linn. Soc.' vol. ii.
1857, p. 7. See also Professor
Oliver in 'Nat. Hist. Review,'
1862, p. 240.

without any opening or fissure." There is only a
single fertile stamen; the style is almost obsolete,
with the three stigmatic surfaces directed to one side.
Both the perfect and cleistogamic flowers produce
seeds.*

The cleistogamic flowers on some of the Malpighiaceæ seem to be more profoundly modified than
those in any of the foregoing genera. According to
A. de Jussieu† they are differently situated from the
perfect flowers; they contain only a single stamen,
instead of 5 or 6; and it is a strange fact that this
particular stamen is not developed in the perfect
flowers of the same species. The style is absent or
rudimentary; and there are only two ovaries instead
of three. Thus these degraded flowers, as Jussieu
remarks, " laugh at our classifications, for the greater
number of the characters proper to the species, to the
genus, to the family, to the class disappear." I may
add that their calyces are not glandular, and as,
according to Kerner,‡ the fluid secreted by such
glands generally serves to protect the flowers from
crawling insects, which steal the nectar without aiding
in their cross-fertilisation, the deficiency of the glands
in the cleistogamic flowers of these plants may perhaps
be accounted for by their not requiring any such
protection.

As the Asclepiadous genus Stapelia is said to produce cleistogamic flowers, the following case may be
worth giving. I have never heard of the perfect flowers
of *Hoya carnosa* setting seeds in this country, but some
capsules were produced in Mr. Farrer's hot-house;

---

* Dr. Kirk, 'Journ. Linn. Soc.
vol. viii. 1864, p. 147.
  † 'Archives du Muséum,' tom.
iii. 1843, pp. 35–38, 82–86, 589, 598.

‡ 'Die Schutzmittel der Blüthen
gegen unberufene Gäste,' 1876,
p. 25.

and the gardener detected that they were the product
of minute bud-like bodies, three or four of which
could sometimes be found on the same umbel with the
perfect flowers. They were quite closed and hardly
thicker than their peduncles. The sepals presented
nothing particular, but internally and alternating
with them, there were five small flattened heart-shaped
papillæ, like rudiments of petals ; but the homological
nature of which appeared doubtful to Mr. Bentham
and Dr. Hooker. No trace of anthers or of stamens
could be detected ; and I knew from having examined
many cleistogamic flowers what to look for. There
were two ovaries, full of ovules, quite open at their
upper ends, with their edges festooned, but with no
trace of a proper stigma. In all these flowers one of
the two ovaries withered and blackened long before
the other. The one perfect capsule, 3½ inches in
length, which was sent me, had likewise been developed from a single carpel. This capsule contained an abundance of plumose seeds, many of which
appeared quite sound, but they did not germinate
when sown at Kew. Therefore the little bud-like
flower which produced this capsule probably was as
destitute of pollen as were those which I examined.

Juncus bufonius and Hordeum.—All the species
hitherto mentioned which produce cleistogamic
flowers are entomophilous ; but four genera, Juncus,
Hordeum, Cryptostachys, and Leersia are anemophilous. *Juncus bufonius* is remarkable* by bearing in
parts of Russia only cleistogamic flowers, which contain three instead of the six anthers found in the
perfect flowers. In the genus *Hordeum* it has been

---

* See Dr. Ascherson's interesting paper in ' Bot. Zeitung,' 1871,
p. 551.

shown by Delpino* that the majority of the flowers are cleistogamic, some of the others expanding and apparently allowing of cross-fertilisation. I hear from Fritz Müller that there is a grass in Southern Brazil, in which the sheath of the uppermost leaf, half a metre in length, envelopes the whole panicle; and this sheath never opens until the self-fertilised seeds are ripe. On the roadside some plants had been cut down, whilst the cleistogamic panicles were developing, and these plants afterwards produced free or unenclosed panicles of small size, bearing perfect flowers.

Leersia oryzoides.—It has long been known that this plant produces cleistogamic flowers, but these were first described with care by M. Duval-Jouve.† I procured plants from a stream near Reigate, and cultivated them for several years in my green-house. The cleistogamic flowers are very small, and usually mature their seeds within the sheaths of the leaves. These flowers are said by Duval-Jouve to be filled by slightly viscid fluid; but this was not the case with several that I opened; but there was a thin film of fluid between the coats of the glumes, and when these were pressed the fluid moved about, giving a singularly deceptive appearance of the whole inside of the flower being thus filled. The stigma is very small and the filaments extremely short; the anthers are less than $\frac{1}{50}$ of an inch in length or about one-third of the length of those in the perfect flowers. One of the three anthers dehisces before the two others. Can this have any relation with the fact that in some other

---

* 'Bollettini del Comizio agrario Parmense.' Marzo e Aprile, 1871. An abstract of this valuable paper is given in 'Bot. Zeitung,' 1871, p. 537. See also Hildebrand on Hordeum, in 'Monatsbericht d. K. Akad. Berlin,' Oct. 1872, p. 760.
† 'Bull. Bot. Soc. de France,' tom. x. 1863, p. 194.

species of Leersia only two stamens are fully developed ?* The anthers shed their pollen on the
stigma; at least in one instance this was clearly the
case, and by tearing open the anthers under water
the grains were easily detached. Towards the apex of
the anther the grains are arranged in a single row and
lower down in two or three rows, so that they could be
counted; and there were about 35 in each cell, or 70
in the whole anther; and this is an astonishingly small
number for an anemophilous plant. The grains have
very delicate coats, are spherical and about $\frac{5}{7000}$ of
an inch (·0181 mm.), whilst those of the perfect flowers
are about $\frac{7}{7000}$ of an inch (·0254 mm.) in diameter.

M. Duval-Jouve states that the panicles very rarely
protrude from their sheaths, but that when this does
happen the flowers expand and exhibit well-developed
ovaries and stigmas, together with full-sized anthers
containing apparently sound pollen; nevertheless such
flowers are invariably quite sterile. Schreiber had previously observed that if a panicle is only half protruded,
this half is sterile, whilst the still included half is
fertile. Some plants which grew in a large tub of
water in my green-house behaved on one occasion in a
very different manner. They protruded two very
large much-branched panicles; but the florets never
opened, though these included fully developed stigmas, and stamens supported on long filaments with
large anthers that dehisced properly. If these florets
had opened for a short time unperceived by me and
had then closed again, the empty anthers would
have been left dangling outside. Nevertheless they
yielded on August 17th an abundance of fine ripe
seeds. Here then we have a near approach to the

---

* Asa Gray, 'Manual of Bot. of United States,' 1856, p. 540.

single case as yet known* of this grass producing in a
state of nature (in Germany) perfect flowers which
yielded a copious supply of fruit. Seeds from the cleistogamic flowers were sent by me to Mr. Scott in
Calcutta, who there cultivated the plants in various
ways, but they never produced perfect flowers.

In Europe *Leersia oryzoides* is the sole representa-
tive of its genus, and Duval-Jouve, after examining
several exotic species, found that it apparently is the
sole one which bears cleistogamic flowers. It ranges
from Persia to North America, and specimens from
Pennsylvania resembled the European ones in their
concealed manner of fructification. There can therefore be little doubt that this plant generally propa-
gates itself throughout an immense area by cleistogamic seeds, and that it can hardly ever be invigorated
by cross-fertilisation. It resembles in this respect
those plants which are now widely spread, though they
increase solely by asexual generation.†

Concluding Remarks on Cleistogamic Flowers.—That
these flowers owe their structure primarily to the
arrested development of perfect ones, we may infer
from such cases as that of the lower rudimentary petal
in Viola being larger than the others, like the lower
lip of the perfect flower,—from a vestige of a spur in
the cleistogamic flowers of Impatiens,—from the ten
stamens of Ononis being united into a tube,—and
other such structures. The same inference may be
drawn from the occurrence, in some instances, on the
same plant of a series of gradations between the
cleistogamic and perfect flowers. But that the former
owe their origin wholly to arrested development is

---

* Dr. Ascherson, ' Bot. Zeitung,'
1864, p. 350.
† I have collected several such
cases in my ' Variation under
Domestication,' ch. xviii.—2nd
edit. vol. ii. p. 153.

by no means the case; for various parts have been
specially modified, so as to aid in the self-fertilisation
of the flowers, and as a protection to the pollen; for
instance, the hook-shaped pistil in Viola and in some
other genera, by which the stigma is brought close
to the fertile anthers,—the rudimentary corolla of
Specularia modified into a perfectly closed tympanum,
and the sheath of Monochoria modified into a closed
sack,—the excessively thin coats of the pollen-grains,
—the anthers not being all equally aborted, and other
such cases.   Moreover Mr. Bennett has shown that
the buds of the cleistogamic and perfect flowers of
Impatiens differ at a very early period of growth.

The degree to which many of the most important
organs in these degraded flowers have been reduced
or even wholly obliterated, is one of their most remarkable peculiarities, reminding us of many parasitic
animals.   In some cases only a single anther is left,
and this contains but few pollen-grains of diminished
size; in other cases the stigma has disappeared,
leaving a simple open passage into the ovarium.   It
is also interesting to note the complete loss of trifling
points in the structure or functions of certain parts,
which though of service to the perfect flowers, are of
none to the cleistogamic; for instance the collecting
hairs on the pistil of Specularia, the glands on the
calyx of the Malpighiaceæ, the nectar-secreting appendages to the lower stamens of Viola, the secretion
of nectar by other parts, the emission of a sweet odour,
and apparently the elasticity of the valves in the
buried capsules of *Viola odorata*.   We here see, as
throughout nature, that as soon as any part or
character becomes superfluous it tends sooner or later
to disappear.

Another peculiarity in these flowers is that the

pollen-grains generally emit their tubes whilst still enclosed within the anthers; but this is not so remarkable a fact as was formerly thought, when the case of Asclepias was alone known.* It is, however, a wonderful sight to behold the tubes directing themselves in a straight line to the stigma, when this is at some little distance from the anthers. As soon as they reach the stigma or the open passage leading into the ovarium, no doubt they penetrate it, guided by the same means, whatever these may be, as in the case of ordinary flowers. I thought that they might be guided by the avoidance of light: some pollen-grains of a willow were therefore immersed in an extremely weak solution of honey, and the vessel was placed so that the light entered only in one direction, laterally or from below or from above, but the long tubes were in each case protruded in every possible direction.

As cleistogamic flowers are completely closed they are necessarily self-fertilised, not to mention the absence of any attraction to insects; and they thus differ widely from the great majority of ordinary flowers. Delpino believes† that cleistogamic flowers have been developed in order to ensure the production of seeds under climatic or other conditions which tend

---

* The case of Asclepias was described by R. Brown. Baillon asserts ('Adansonia,' tom. ii. 1862, p. 58) that with many plants the tubes are emitted from pollen-grains which have not come into contact with the stigma; and that they may be seen advancing horizontally through the air towards the stigma. I have observed the emission of the tubes from the pollen-masses whilst still within the anthers, in three widely distinct Orchidean genera namely Aceras, Malaxis, and Neottia: see 'The Various Contrivances by which Orchids are Fertilised,' 2nd edit. p. 258.

† 'Sull' Opera la Distribuzione dei Sessi nelle Piante,' 1867, p. 30.

to prevent the fertilisation of the perfect flowers. I do not doubt that this holds good to a certain limited extent, but the production of a large supply of seeds with little consumption of nutrient matter or expenditure of vital force is probably a far more efficient motive power. The whole flower is much reduced in size; but what is much more important, an extremely small quantity of pollen has to be formed, as none is lost through the action of insects or the weather; and pollen contains much nitrogen and phosphorus. Von Mohl estimated that a single cleistogamic anther-cell of *Oxalis aceto-sella* contained from one to two dozen pollen-grains; we will say 20, and if so the whole flower can have produced at most 400 grains; with Impatiens the whole number may be estimated in the same manner at 250; with Leersia at 210; and with *Viola nana* at only 100. These figures are wonderfully low compared with the 243,600 pollen-grains produced by a flower of Leontodon, the 4,863 by an Hibiscus, or the 3,654,000 by a Pæony.* We thus see that cleistogamic flowers produce seeds with a wonderfully small expenditure of pollen; and they produce as a general rule quite as many seeds as the perfect flowers.

That the production of a large number of seeds is necessary or beneficial to many plants needs no evidence. So of course is their preservation before they are ready for germination; and it is one of the many remarkable peculiarities of the plants which bear cleistogamic flowers, that an incomparably larger proportion of them than of ordinary plants bury their young ovaries in the ground;—an action which it may be presumed serves to protect them from being

---

* The authorities for these statements are given in my ' Effects of Cross and Self-Fertilisation,' p. 376.

devoured by birds or other enemies. But this advantage is accompanied by the loss of the power of wide dissemination. No less than eight of the genera in the list at the beginning of this chapter include species which act in this manner, namely, several kinds of Viola, Oxalis, Vandellia, Linaria, Commelina, and at least three genera of Leguminosæ. The seeds also of Leersia, though not buried, are concealed in the most perfect manner within the sheaths of the leaves. Cleistogamic flowers possess great facilities for burying their young ovaries or capsules, owing to their small size, pointed shape, closed condition and the absence of a corolla; and we can thus understand how it is that so many of them have acquired this curious habit.

It has already been shown that in about 32 out of the 55 genera in the list just referred to, the perfect flowers are irregular; and this implies that they have been specially adapted for fertilisation by insects. Moreover three of the genera with regular flowers are adapted by other means for the same end. Flowers thus constructed are liable during certain seasons to be imperfectly fertilised, namely, when the proper insects are scarce; and it is difficult to avoid the belief that the production of cleistogamic flowers, which ensures under all circumstances a full supply of seed, has been in part determined by the perfect flowers being liable to fail in their fertilisation. But if this determining cause be a real one, it must be of subordinate importance, as four of the genera in the list are fertilised by the wind; and there seems no reason why their perfect flowers should fail to be fertilised more frequently than those in any other anemophilous genus. In contrast with what we here see with respect to the large proportion of the perfect flowers being irregular, one genus alone out of the 38 heterostyled genera described in the previous chapters bears such flowers; yet all these genera are absolutely dependent on insects for their legitimate fertilisation. I know not how to account for this difference in the proportion of the plants bearing regular and irregular flowers in the two classes, unless it be that the heterostyled flowers are already so well adapted for cross-fertilisation, through the position of their stamens and pistils and the difference in power of their two or three kinds of pollen, that any additional adaptation, namely, through the flowers being made irregular, has been rendered superfluous.

Although cleistogamic flowers never fail to yield a large number of seeds, yet the plants bearing them usually produce perfect flowers, either simultaneously or more commonly at a different period; and these are adapted for or admit of cross-fertilisation. From the cases given of the two Indian species of Viola, which produced in this country during several years only cleistogamic flowers, and of the numerous plants of Vandellia and of some plants of Ononis which behaved during one whole season in the same manner, it appears rash to infer from such cases as that of *Salvia cleistogama* not having produced perfect flowers during five years in Germany,* and of an Aspicarpa not having done so during several years in Paris, that these plants would not bear perfect flowers in their native homes. Von Mohl and several other botanists have repeatedly insisted that as a general rule the perfect flowers produced by cleistogamic plants are sterile; but it has been shown under the head of the several species that this is not the case. The perfect

---

* Dr. Ascherson, ' Bot. Zeit.' 1871, p. 555.

flowers of Viola are indeed sterile unless they are visited by bees; but when thus visited they yield the full number of seeds. As far as I have been able to discover there is only one absolute exception to the rule that the perfect flowers are fertile, namely, that of Voandzeia; and in this case we should remember that cultivation often affects injuriously the reproductive organs. Although the perfect flowers of Leersia sometimes yield seeds, yet this occurs so rarely, as far as hitherto observed, that it practically forms a second exception to the rule.

As cleistogamic flowers are invariably fertilised, and as they are produced in large numbers, they yield altogether a much larger supply of seeds than do the perfect flowers on the same plant. But the latter flowers will occasionally be cross-fertilised, and their offspring will thus be invigorated, as we may infer from a wide-spread analogy. But of such invigoration I have only a small amount of direct evidence: two crossed seedlings of *Ononis minutissima* were put into competition with two seedlings raised from cleistogamic flowers; they were at first all of equal height; the crossed were then slightly beaten; but on the following year they showed the usual superiority of their class, and were to the self-fertilised plants of cleistogamic origin as 100 to 88 in mean height. With Vandellia twenty crossed plants exceeded in height twenty plants raised from cleistogamic seeds only by a little, namely, in the ratio of 100 to 94.

It is a natural inquiry how so many plants belonging to various very distinct families first came to have the development of their flowers arrested, so as ultimately to become cleistogamic. That a passage from the one state to the other is far from difficult is shown by the many recorded cases of gradations between the two states on the same plant, in Viola, Oxalis, Biophytum, Campanula, &c. In the several species of Viola the various parts of the flowers have also been modified in very different degrees. Those plants which in their own country produce flowers of full or nearly full size, but never expand (as with Thelymitra), and yet set fruit, might easily be rendered cleistogamic. *Lathyrus nissolia* seems to be in an incipient transitional state, as does *Drosera Anglica*, the flowers of which are not perfectly closed. There is good evidence that flowers sometimes fail to expand and are somewhat reduced in size, owing to exposure to unfavourable conditions, but still retain their fertility unimpaired. Linnæus observed in 1753 that the flowers on several plants brought from Spain and grown at Upsala did not show any corolla and yet produced seeds. Asa Gray has seen flowers on exotic plants in the Northern United States which never expanded and yet fruited. With certain English plants, which bear flowers during nearly the whole year, Mr. Bennett found that those produced during the winter season were fertilised in the bud; whilst with other species having fixed times for flowering, but "which had been tempted by a mild January to put forth a few wretched flowers," no pollen was discharged from the anthers, and no seed was formed. The flowers of *Lysimachia vulgaris* if fully exposed to the sun expand properly, while those growing in shady ditches have smaller corollas which open only slightly; and these two forms graduate into one another in intermediate stations. Herr Bouché's observations are of especial interest, for he shows that both temperature and the amount of light affect the size of the corolla; and he gives measurements proving that with some plants the corolla is diminished by the increasing cold and darkness of the changing season, whilst with others it is diminished by the increasing heat and light.*

The belief that the first step towards flowers being rendered cleistogamic was due to the conditions to which they were exposed, is supported by the fact of various plants belonging to this class either not producing their cleistogamic flowers under certain conditions, or, on the other hand, producing them to the complete exclusion of the perfect ones. Thus some species of Viola do not bear cleistogamic flowers when growing on the lowlands or in certain districts. Other plants when cultivated have failed to produce perfect flowers during several successive years; and this is the case with *Juncus bufonius* in its native land of Russia. Cleistogamic flowers are produced by some species late and by others early in the season; and this agrees with the view that the first step towards their development was due to climate; though the periods at which the two sorts of flowers now appear must since have become much more distinctly defined. We do not know whether too low or too high a temperature or the amount of light acts in a direct manner on the size of the corolla, or indirectly through the male organs being first affected. However this may be, if a plant were prevented either early or late in the season from fully expanding its corolla, with some reduction in its size, but with no loss of the power of self-fertilisation, then natural selection might well complete the work and

---

* For the statement by Linnæus, see Mohl in 'Bot. Zeitung,' 1863, p. 327. Asa Gray, 'American Journal of Science,' 2nd series, vol. xxxix. 1865, p. 105. Bennett in 'Nature,' Nov. 1869, p. 11. The Rev. G. Henslow also says ('Gardener's Chronicle,' 1877, p. 271: also 'Nature,' Oct. 19, 1876, p. 543) " that when the autumn draws on, and habitually in winter for such of our wild flowers as blossom at that season," the flowers are self-fertilised. On Lysimachia, H. Müller, 'Nature,' Sept. 1873, p. 433. Bouché, 'Sitzungsbericht der Gesell. Naturforsch. Freunde,' Oct. 1874, p. 90.

render it strictly cleistogamic. The various organs would also, it is probable, be modified by the peculiar conditions to which they are subjected within a completely closed flower; also by the principle of correlated growth, and by the tendency in all reduced organs finally to disappear. The result would be the production of cleistogamic flowers such as we now see them; and these are admirably fitted to yield a copious supply of seed at a wonderfully small cost to the plant.

I will now sum up very briefly the chief conclusions which seem to follow from the observations given in this volume. Cleistogamic flowers afford, as just stated, an abundant supply of seeds with little expenditure; and we can hardly doubt that they have had their structure modified and degraded for this special purpose; perfect flowers being still almost always produced so as to allow of occasional cross-fertilisation. Hermaphrodite plants have often been rendered monœcious, diœcious or polygamous; but as the separation of the sexes would have been injurious, had not pollen been already transported habitually by insects or by the wind from flower to flower, we may assume that the process of separation did not commence and was not completed for the sake of the advantages to be gained from cross-fertilisation. The sole motive for the separation of the sexes which occurs to me, is that the production of a great number of seeds might become superfluous to a plant under changed conditions of life; and it might then be highly beneficial to it that the same flower or the same individual should not have its vital powers taxed, under the struggle for life to which all organisms are subjected, by producing both pollen and seeds. With respect to the plants belonging to the gyno-diœcious sub-class, or those which co-exist as hermaphrodites and females, it has been proved that they yield a much larger supply of seed than they would have done if they had all remained hermaphrodites; and we may feel sure from the large number of seeds produced by many plants that such production is often necessary or advantageous. It is therefore probable that the two forms in this sub-class have been separated or developed for this special end.

Various hermaphrodite plants have become heterostyled, and now exist under two or three forms; and we may confidently believe that this has been effected in order that cross-fertilisation should be assured. For the full and legitimate fertilisation of these plants pollen from the one form must be applied to the stigma of another. If the sexual elements belonging to the same form are united the union is an illegitimate one and more or less sterile. With dimorphic species two illegitimate unions, and with trimorphic species twelve are possible. There is reason to believe that the sterility of these unions has not been specially acquired, but follows as an incidental result from the sexual elements of the two or three forms having been adapted to act on one another in a particular manner, so that any other kind of union is inefficient, like that between distinct species. Another and still more remarkable incidental result is that the seedlings from an illegitimate union are often dwarfed and more or less or completely barren, like hybrids from the union of two widely distinct species.

INDEX.

LONDON: PRINTED BY WILLIAM CLOWES AND SONS, STAMFORD STREET
AND CHARING CROSS.

Printed in the United States
By Bookmasters